T0329582

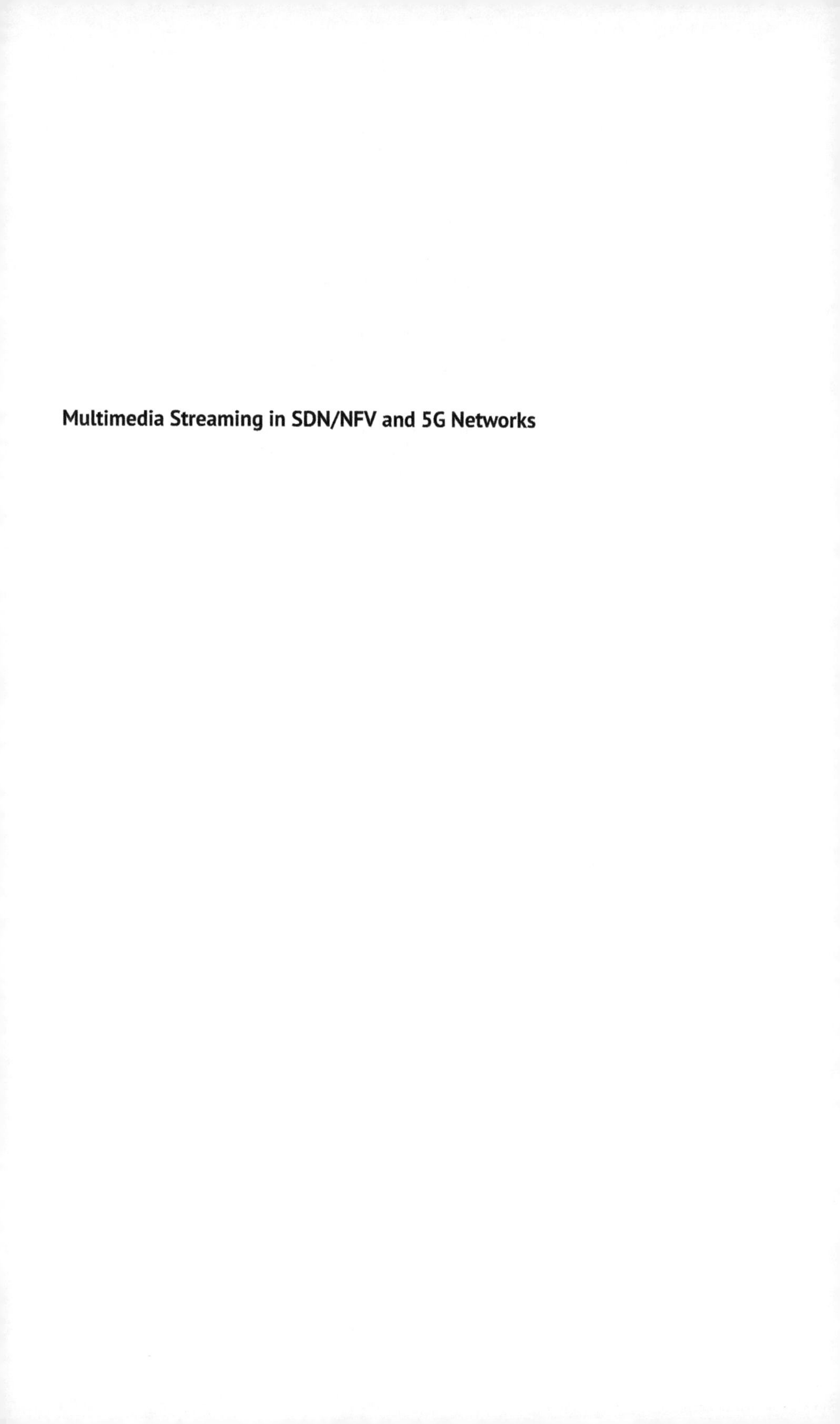

Multimedia Streaming in SDN/NFV and 5G Networks

IEEE Press
445 Hoes Lane
Piscataway, NJ 08854

Multimedia Streaming in SDN/NFV and 5G Networks

Machine Learning for Managing Big Data Streaming

Alcardo Barakabitze and Andrew Hines
University College Dublin (UCD), Dublin, Ireland

Published by John Wiley & Sons, Inc., Hoboken, New Jersey.
Published simultaneously in Canada.

Library of Congress Cataloging-in-Publication Data Applied for:

Hardback ISBN: 9781119800798

Cover Design: Wiley
Cover Image: © Sergey Nivens/Shutterstock

Set in 9.5/12.5pt STIXTwoText by Straive, Chennai, India

Contents

Biography of Authors

Alcardo Barakabitze

Dr. Alcardo Barakabitze is a former MSCA Fellow in the School of Computing, Dublin City University, Ireland. Dr. Barakabitze completed his PhD in Computing and Communications from the School of Science and Engineering at the University of Plymouth, UK. His PhD thesis was titled "Quality of Experience (QoE) control and management of multimedia services in software defined and virtualized networks". He received the degree in Computer Science with Honours from the University of Dar es Salaam, Tanzania in 2010 and Master Degree of Electronics and Communication Engineering with first class from Chongqing University, PR China, in May 2015. He has worked as Marie Curie Fellow (2015–2018) in MSCA ITN QoE-Net project. He was also a Postdoctoral Research Fellow (2019–2020) at Qxlab in the School of Computer Science, University College Dublin under the Government of Ireland Postdoctoral Fellowship (Irish Research Council).

Dr. Barakabitze is recognised as the 2015 outstanding International Graduate Student of Chongqing University, China due to his excellent performance. He was a visiting researcher in the Department of Electrical and Electronics Engineering, University of Cagliari, Italy and the ITU-T-Standardization Department in 2016 and 2017 respectively. He is actively involved in both national and international projects from the EU Horizon 2020, and Horizon Europe – 2027 programme. He is an active member of the standardization activities of the ITU-T Study Group 12 and 13 focusing on network performance, QoE and future 6G and 2030 networks. Dr. Barakabitze has more than 40 publications in peer-reviewed conferences and journals with a total of 1041 Google Scholar citations and H-index of 17. Barakabitze served as session chair of Future Internet and NGN Architectures during the IEEE Communication Conference in Kansas City, USA. He was also the Keynote and Panel Chairs at the International Young Researcher Summit on Quality of Experience in Emerging Multimedia Services (QEEMS 2017), which was held from May 29–30, 2017 in Erfurt, Germany. Dr. Barakabitze is a Reviewer for various journals including the IEEE Transactions in Network and Service

Management, IEEE ACCESS, Computer Networks and many others. Dr. Barakabitze serves on technical program committees of leading conferences focusing on his research areas. His research interests are 5G, Quality of Experience (QoE), network management, video streaming services, SDN and NFV.

Andrew Hines
Andrew Hines is an Assistant Professor with the School of Computer Science, University College Dublin, Ireland where he served as the Director of Research, Impact and Innovation. He leads the QxLab research group with active research in multimedia signal processing as well as Quality of Experience for 5G, SDN and network management. He is an investigator in the Science Foundation Ireland Connect, Insight and Adapt research centres. He was awarded the US-Ireland Research Innovation award in 2018 by the Royal Irish Academy and American Chamber of Commerce for work developing Multimedia Quality of Experience Models. He is a member of the 2022–2026 Royal Irish Academy Multidisciplinary Committee (Engineering and Computer Science). Dr Hines is a senior member of the IEEE and a member of the Audio Engineering Society. As a leading expert in Quality of Experience for media technology, he represented Ireland on the management committee of the European COST action expert group on Quality of Experience, Qualinet.

Abstract

The exponential growth of video streaming services (e.g. YouTube and Mobile TV) on smart devices has triggered and introduced new revenue potential for telecom operators and service providers. The future networks such as 5G and 6G are shifting toward the cloudification of the network resources via Software Defined Networks (SDNs), Network Function Virtualization (NFV), and Multi-Access Edge/Cloud Computing, network hypervisors, virtual machines, and containers. This will equip Internet Service Providers (ISPs) with cutting-edge technologies to provide service customization during service delivery and to offer Quality of Experience (QoE) which meets customers' needs via intelligent QoE control and management approaches. This book provides a high-level description of QoE control and management solutions for multimedia services in future softwarized and virtualized 5G networks. The QoE control and management integrates QoE modeling, monitoring, and optimization at different points in the network. The book starts with an introduction to 5G networks along with 5G service quality and business requirements. The book provides a comprehensive description of HTTP2/3 Adaptive Streaming (HAS) solutions as the dominant technique for streaming high-definition videos (4K/8K/3D) over the future Internet architecture. The book provides a description of network softwarization and virtualization technologies leveraging SDN, NFV, MEC, Fog/Cloud Computing as important elements for managing multimedia services in future networks such as 5G/6G. It also provides the past and ongoing key research projects, standardization activities, and use cases related to SDN, NFV, MEC, Fog/Cloud, and emerging applications (cloud video streaming gaming, immersive virtual reality [VR] and augmented reality [AR], mulsemedia, and 360° immersive video).

Moreover, the book provides the QoE control and management techniques using softwarized and virtualized networks. Furthermore, this book provides the 5G network slicing in future softwarized networks which is a key element for achieving QoE-oriented sharing/slicing of resources. The book will provide different industrial initiatives and projects that are pushing forward the adoption of network softwarization and virtualization in accelerating 5G network slicing. A comparison of various 5G architectural approaches in terms of practical implementations, technology adoptions, and deployment strategies will be presented in the book. In addition, a discussion on various open-source orchestrators and proof of concepts representing industrial contribution will be given in the book. This will go along with the management and orchestration of network slices in multi-tenant or multi-operators domains. In addition to that, the book provides a multimedia QoE-driven services delivery toward 6G and 2030 networks. Some of the key technologies and services highlighted in this chapter include: holographic and future media communications, human-centric services and 3D volumetric video streaming, innovative network architectures, pervasive AI, ML, and big data analytics, new protocol stack and 3D network architectures. The future of video streaming industry towards 2030 and new paradigms of internetworking for 2030 and beyond systems are also highlighted in the book. The last chapter provides future challenges for managing multimedia services focusing on the following important areas: emerging multimedia services, QoE-oriented business models in future softwarized network, intelligent QoE-based big data strategies in SDN/NFV, scalability, resilience, and optimization in SDN and NFV, multimedia communications in the Internet of Things AQ1 (IoTs), OTT-ISP collaborative service management in softwarized networks. The chapter provides various opportunities that need extensive research from the academia and industry regarding QoE-driven virtualized multimedia 3D services delivery schemes over 6G architecture, 3D cloud/edge models for multimedia services and elastic 3D service customization via network intelligentization and QoE management and orchestration challenges, over 6G networks.

List of Figures

List of Tables

Preface

This book expands on his published papers in the IEEE Communication Surveys and Tutorials and Elsevier Computer Networks journals. The transitional narrative of the material from these articles were added at the start of every chapter and section. The intention is to give a compact, highly motivated introduction to the central questions that arise in the study of QoE management of multimedia services in future 5G networks. The book provides a high-level description of QoE control and management solutions for multimedia services in future softwarized and Virtualized 5G networks. The book covers network softwarization and virtualization technologies including SDN, NFV, Multi-Access Edge Computing (MEC), Fog/Cloud Computing as important elements for managing multimedia services in future networks such as 5G/6G and beyond. The book further examines the end-to-end QoE control and management solutions in softwarized and virtualized networks including: (i) QoE-aware/driven strategies using SDN or/and NFV, (ii) QoE-aware/driven approaches for adaptive streaming over emerging architectures such as MEC, cloud/fog computing, and Information-Centric Networking, and (iii) QoE measurements in new domains including AR/VR, mulsemedia and video gaming applications. The final part of this book discusses future challenges and research directions/recommendations in: (i) Emerging multimedia services and applications, (ii) 5G network management and orchestration, (iii) network slicing and collaborative service management of multimedia services in softwarized networks, and (iv) QoE-oriented business models. The book is intended for researchers, engineers from academia and industry working in the area of multimedia services and telecommunication communications, networking and QoE optimization and control and management. The book is also intended for application scientists from the industry for gaining important knowledge of QoE management of multimedia services in future 5G and beyond networks using network softwarization technologies (SDN and NFV) and other emerging architectures (MEC, ICN, Cloud/Edge computing). In particular, the book will be beneficial to undergraduate students studying for a degree in computer science, communication engineering and telecommunication systems for both taught courses and their project work. In particular, the book will be also useful for postgraduate students for a PhD or master's degree in the above courses. I believe that, for undergraduate and postgraduate students, this book will be a valuable source of information for fundamental topics regarding video streaming QoE, multimedia services and applications, 5G networks, future Internet architecture, and softwarized and virtualized networks leveraging SDN and NFV. I also believe that the book will be used as a good vehicle for self-learning and teaching text tools for application engineers, undergraduate and graduate students and university academics. Readers of this book are assumed to have some basic knowledge of computer networking, and an interest in SDN, NFV, cloud computing, multimedia services, video quality optimization, and QoE services delivery to end-users. The book will also be a valuable resource for experts working in different fields such as patent attorneys and patent examiners.

Dr. Alcardo Barakabitze completed his PhD in Computing and Communications from the University of Plymouth, UK, in 2019. He received the degree in Computer Science with honors from the University of Dar es Salaam, Tanzania, in 2010 and master's degree of Electronics and Communication Engineering with first class from Chongqing University, PR China, in May 2015. He is a former MSCA Fellow (October 2020 – September 20228) in the School of Computing, Dublin City University, Ireland. In June 2020, Dr. Barakabitze delivered a commissioned four-day training workshop on the topics in this book proposal for a group of patent examiners based in the European Patent Office, Geneva.

Since October 2019 to September 2020, he has served as a postdoctoral researcher in the QxLab research group, School of Computer Science, University College Dublin (UCD). He has worked as Marie Curie Fellow (October 2015–July 2019) in MSCA ITN QoE-Net. Dr. Barakabitze was recognized as the 2015 outstanding International Graduate Student of Chongqing University, China, due to his excellent performance. He was awarded the BEST Male ICT Researcher of the Year 2021 by the ICT Commission of Tanzania. Dr. Barakabitze is also recognized as the BEST Young Researcher of the Year 2021, an award from the SUA Research and Innovation Competitions. He was a visiting researcher in the Department of Electrical and Electronics Engineering, University of Cagliari, Italy, and the ITU-T-Standardization Department in 2016 and 2017, respectively. Dr. Barakabitze has participated in the ITU-T standardization process and made significant contributions since 2016 in the area of Software Defined Networking (SDN)/Network Function Virtualization (NFV), future network performance, Quality of Service (QoS), and Quality of Experience (QoE). Dr. Barakabitze has served as session chair of Future Internet and NGN Architectures during the 2018 IEEE Communication Conference (ICC) in Kansas City, USA. He was also the Keynote and Panel Chairs at the International Young Researcher Summit on Quality of Experience in Emerging Multimedia Services (QEEMS 2017), that was held from May 29 to 30, 2017 in Erfurt, Germany. He has numerous publications in international peer-reviewed conferences and journals. He is a reviewer for various journals and serves on technical program committees of leading conferences focusing on his research areas. His research interests are 5G/6G, QoE, network management, video streaming services, SDN, and NFV.

Acknowledgments

Writing a book is harder than I thought and more rewarding than I could have ever imagined. This work is the culmination of a journey that started in October 2020. Completing this book has been a truly life-changing experience for me, and it would not have been possible to do without the support and guidance that I received from many people. First and foremost, I would like to express my sincere gratitude to my supervisors of Postdoc (Prof. Andrew and Hines and Ray Walshe) for the continuous support, patience, motivation, and immense knowledge they gave to me since I started my Postdoc study at UCD and DCU. Their professional guidance helped me in all the time of research and writing of this book. I appreciate all the contributions of time, encouragement, and ideas to make the experience of writing this book productive and stimulating. The joy and enthusiasm they have for their research will remain to be contagious and motivational for me throughout the journey of my research career. It has been an enormous privilege to learn from their expertise and leadership, and I am also thankful for the excellent example they have provided to me of working very hard. My numerous quality chapters that I was able to publish during my two years of writing this book would not have been possible without them and for that I am truly grateful. I am also deeply in debt to my project managers, Mr. Paulo Soncini and Dr. Rob Brennan, who always had a door open when I needed someone to discuss and get feedback from. Their insights and comments have shaped this book. A very special thanks to them for their advice and support during my two years of doing my Postdocs. I have hugely benefited from the collaborations and discussions with them. I collectively thank my project supervisory team for inspirational guidance, support, and constructive discussions with me regarding the chapters. I am truly grateful to Dr. Andrew Hines for helping me in whatever way he could during this challenging period. I would also like to thank all the reviewers who reviewed my book. Without their constructive comments and suggestions, I would not have been able to raise the standard of the book chapters.

A particular mention also goes to all members of QxLab Research group at the University College Dublin (UCD), and the Department of Informatics and Information Technologies at Sokoine University of Agriculture (SUA), Tanzania, and Huawei Canada for any kind of help, support, and constructive discussions they gave to me through Zoom. It has been my great pleasure to have been working with these teams. I thank all of my fellow lab mates and coworkers from different universities, industry, and IT companies for the stimulating discussions and for all the fun moments we have had together in the last two years of completing this book. It has really been a privilege to be part of such an amazing team. I gratefully acknowledge the funding received toward my Postdoc from the European Union's Horizon 2020 Research and Innovation Fellowship under the Marie Skłodowska-Curie-Innovative Training Networks (MSCA ITN), ELITE-S Fellowship Programme: A Marie Skłodowska-Curie COFUND Action for intersectoral training, career development and mobility. Infinite thanks to my parents, Alex Barakabitze and Leonia Francis Nzika, my sisters and brothers, Josaphine, Alfred, Abely, and Bibiana, for all the support and unconditional love over the years. I would also like to thank my wife, Lilian Alcardo Barakabitze, for her encouragement, support, and unconditional love during this amazing journey of completing the book. Lastly, this book is dedicated to our beautiful three children, Alvin, Niah, and Doreen, for their motivation, support, endless love, and for bearing with me in my Postdocs journey. Not only have they given me the strength, stability, and motivation to finish this work, they have also made my life brighter and lighter with their love.

List of Acronyms

5G	Fifth Generation
ACTN	Abstraction and Control of Traffic Engineered Networks
B2B	Business-to-Business
B2C	Business-to-Customer
BSS	Business Support System
BSSO	Business Service Slice Orchestrator
CAPEX	Capital Expenditure
CC	Cloud Computing
CDNs	CDNs Content Distribution Networks
C-RAN	Cloud RAN
D2D	Device-to-Device
DHCP	Dynamic Host Configuration Protocol
DSSO	Domain-Specific Slice Orchestration
EC2	Elastic Compute Cloud
ELA	Experience Level Agreement
ETSI	European Telecommunication Standard Institute
FoC	Fog Computing
IRTF	Internet Research Task Force
ISPs	Internet Service Providers
ITU	International Telecommunication Union
KPR	Key Performance Requirements
KQIs	Key Quality Indicators
LAN	Local Area Network
LSDC	Lightweight Slice Defined Cloud
M2M	Machine-to-Machine
MANO	Management and Orchestration
MdO	Multi-domain Orchestrator
MDSO	Multi-Domain Slice Orchestrator
MEC	Multi-Access Edge Computing

MIoTs	Massive Internet of Things
MO	Management and Orchestration
MTC	Machine Type Communications
MTCP	Mobile Transport and Computing Platform
NAT	Network Address Translation
NFs	Network Functions
NFV	Network Function Virtualization
NFVI-PoP	NFVI Point of Presence
NFVO	Network Functions Virtualization Orchestrator
NGN	Next Generation Networks
ONF	Open Network Foundation
OPEX	OPerational EXpenditure
OSS	Operations Support Systems
PGW	Packet Data Network Gateway
PoP	Point of Presence
QoBiz	Quality of Business
RAN	Radio Access Network
RLC	Radio Link Control
RRM	Radio Resource Management
SaaS	Software as a Service
SDMC	Software-Defined Mobile Network Control
SDMO	Software-Defined Mobile network Orchestration
SFC	Service Function Chaining
SGW	Service Gateway
SLAs	Service Level Agreements
SRO	Slice Resource Orchestrator
SBS	Service Broker Stratum
SDO	Standard Developing Organizations
TN	Transport Networks
TOSCA	Topology and Orchestration Specification for Cloud Applications
USDL	Universal Service Definition Language
VMN	Virtual Mobile Networks
VMS	Virtual Machines
VNF-FGs	VNF Forwarding Graphs
VNFs	Virtual Network Functions
VPN	Virtual Private Networks
VR/AR	Virtual/Augmented Reality
WWRF	Wireless World Research Forum
XCI	Xhaul Control Infrastructure
ZOOM	Zero-time Orchestration, Operations and Management

1

5G Networks

From 1G to 4G, mobile communication has been for many years constantly changing our behavior, our communication experience in audio/video, and our lifestyle in general. With 5G arriving, our society will change radically through the realization of the Internet of Everything (IoE) where connected sensors will enable: connected robots for manufacturing and on Industry 4.0 revolution; connected sensors for smart city and connected smart homes. The requirements for ultra-low latency and ultra-high reliability in 5G networks is a game-changer that is going to take the automotive industry from the assisted-driving to connected cars. The 5G network is about enabling new services and applications, connecting new industries, and empowering new user experiences. 5G will connect people and things across a diverse set of other vertical segments including media and entertainment such as immersive and interactive media, collaborative video gaming andAugmented Reality (AR) or Virtual Reality (VR).

This chapter provides an introduction to 5G. It begins with some history of the progress through generations 1–5 of mobile network communications. Next it explains the motivations for 5G and the service and business drivers. It also discusses the emerging future services and applications that will require the capabilities of 5G and beyond to 6G. The chapter introduces a lot of network technology acronyms and terminology that is commonly used in the literature regarding 5G. For those familiar with 5G, there is also some discussion of the 5G standardization activities that are underway.

1.1 History of Mobile Communication Systems

Over the past decades, we have witnessed the tremendous growth of the fixed and wireless industry. Indeed, the mobile communication systems have evolved from a purely analog (1G), and limited to voice communications to 4G digital multimedia systems. The 2G mobile systems support full-duplex communication

Multimedia Streaming in SDN/NFV and 5G Networks: Machine Learning for Managing Big Data Streaming, First Edition. Alcardo Barakabitze and Andrew Hines.

and enable services such as picture messages, text messages, and Multimedia Messaging Service (MMS). Driven by the advancement in the Internet and IP network technology, 3G systems appeared in 2000 to fulfil users' ever increasing demands for data and service quality. 3G offers dedicated digital networks that are used for delivering broadband/multimedia services (e.g. TV streaming, mobile TV, phone calls, video conferencing, and 3D gaming) with an increased bandwidth and data transfer rates. 4G was necessary to meet the demands for multimedia services and applications (3D, HDTV content, Digital Video Broadcasting [DVB]) that require high bandwidth on sophisticated user platforms such as tablets and smartphones. 4G systems provide true wireless broadband services and deliver to its customer's multimedia services with good quality. However, many use cases and new applications with diverse requirements have emerged over the past years. These use cases and applications include: Internet of Things (IoTs); Internet of Vehicles (IoV); AR/VR; Device to Device (D2D) and Machine to Machine (M2M) communications and Financial Technology (FinTech), etc. Figure 1.1 shows the network evolution toward 5G systems and beyond.

It is unclear whether the current 4G LTE cellular systems can support the enormous rapid growth of data usage and device connectivity. For example, IoT and D2D/M2M communications in 5G systems would support tens of thousands of connected smart devices in a single cell while the current 4G LTE network can support up to 600 RCC-connected users per cell. Both industry and academia are embracing 5G as the future network capable of supporting different verticals and use cases consisting of different service requirements.

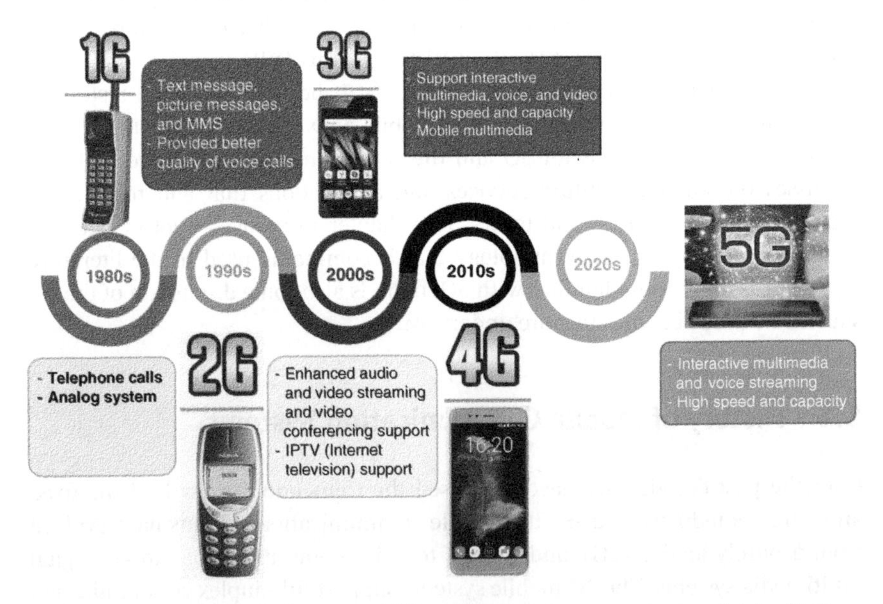

Figure 1.1 Network evolution toward 5G and beyond. Source: sarayut_sy/Adobe Stock.

1.2 5G: Vision and Motivation

With the increasing number of new applications beyond personal communications, mobile devices, the exponential growth of mobile video services (e.g. YouTube and Mobile TV) on smart devices and the advances in the IoT have triggered global initiatives toward developing the 5G mobile/wireless communication systems [1–3]. The Cisco Visual Networking Index (VNI) Forecast [4] predicts that four-fifths of the world's Internet traffic will be IP video traffic by 2023, a ninefold increase from 2018. Mobile video traffic alone will account for 78% of the global mobile data traffic. The traffic growth rates of TVs, tablets, smartphones, and M2M modules will be 21%, 29%, 49%, and 49%, respectively. The traffic for VR/AR will increase at a Compound Annual Growth Rate (CAGR) of 82% between 2018 and 2023. 5G devices and connections will be over 10% of global mobile devices and connections by 2023. Global mobile devices will grow from 8.8 billion in 2018 to 13.1 billion by 2023 – 1.4 billion of those will be 5G capable [4].

Moreover, three-quarters of all devices connected to the mobile network will be smart devices by 2023, generating 99% of all mobile data traffic globally. 5G networks have been driven by the increasing number of smart devices (e.g. tablets and smartphones) and the growing number of bandwidth-hungry mobile applications (e.g. live video streaming, online video gaming) which demand higher spectral efficiency than that of 4G LTE systems [4]. The increasing demand of high-quality services by customers from service providers and mobile network operators and a well-connected society context (e.g. smart grid and smart cities, critical infrastructure systems) are also among the factors that have triggered the development of 5G systems. The 5G market drivers that would enable the global economic output of US12.3 trillion by 2035 include the needs for AR/VR, rich media services (e.g. real-time video gaming, 4K/8K/12/3D video, 360° video, live video broadcasting) and applications in smart cities, education, entertainment, industrial, and public safety [5] (Figure 1.2).

To support these new use cases, emerging multimedia services and applications, 5G and beyond systems should be able to support and deliver as much as 1000 times capacity compared with the current 4G LTE systems [2, 6]. The 5G Key Performance or Quality Indicators (KPIs/KQIs) include: 10–100 times higher user data rates, better, ubiquitous and almost 100% coverage for "anytime anywhere" connectivity and above 90% energy savings. In addition, 5G should also provide an End-to-End (E2E) latency of less than 1 ms, an aggregate service reliability and availability of 99.999%, and lowered electro-magnetic field levels compared with 4G LTE [2, 7]. 5G has been recognized by industry and academia as the game-changer that will enable new video streaming services and applications, connecting new vertical industries with a diverse set of performance and service requirements and empowering new user experiences. 5G will enable a connected society with massive devices connectivity and support intelligent transportation

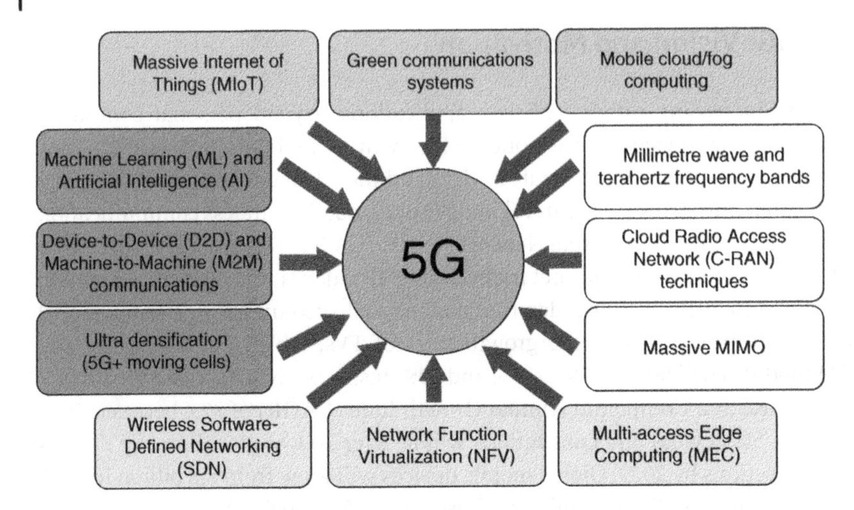

Figure 1.2 The 12 key enabling technologies in 5G networks.

systems (e.g. autonomous and automated driving). Meeting performance targets such as higher data rate transmission, higher capacity, lower E2E latency, lower cost, and user satisfaction measured through Quality of Experience (QoE) for delivered services is key for the success of 5G. The 5G theme has attracted researchers and engineers for the past years. The debates, discussions and preliminary questions regarding the 2020 network and beyond include: (i) What the 5G network will be? [3], (ii) What are the requirements and future technological advancements for 5G networks? [2], (iii) What is the autonomic network architecture that can accommodate and support the emerging services and applications as well as various technologies to address the 5G challenges?, (iv) What are the novel solutions that can incorporate the 5G network driving principles (e.g. seamless mobility, programmability, flexibility): (v) How to implement the 5G vision of network/infrastructure/resource sharing/slicing and support dynamic multi-service, multi-tenancy across network softwarization technologies? (vi) How to perform dynamic control, orchestration of network resources as well as service customization of network slices (enabled by NFV principles) in 5G systems? (vii) How and to what extent can future 5G network management be automated to ensure that ISPs, MNOs, and OTT providers meet customer's service requirements and Experience Level Agreement (ELAs)[1] in the cloud/fog/heterogeneous-native supported softwarized networks [9]?

---

1 ELAs: Indicates a QoE-enabled counter piece to traditional QoS-based Service Level Agreements (SLA) that conveys the performance of the service in terms of QoE. The ELAs establish a common understanding of an end-user experience on the quality levels whiling using the service [8].

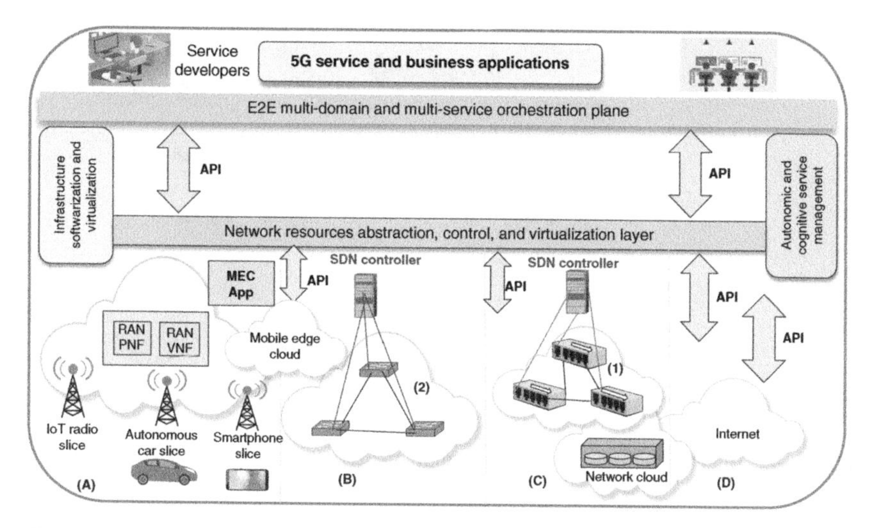

Figure 1.3 Software network technologies in 5G architecture. A indicates RAN; B = transport networks; C = core networks; and D represents the Internet.

Different stakeholders from the research community have expressed their 5G vision for future 2020 communication systems and beyond as shown in Figure 1.3. One of the disruptive concepts that could provide answers to these questions and realize the 5G vision is network softwarization and slicing [10, 11]. For example, future multimedia services such as 4K/8K/12K will be managed and delivered with excellent QoE to end-users in 5G using a new paradigm of network softwarization and virtualization that leverages cutting-edge technologies such as Software-Defined Networking (SDN), Network Function Virtualization (NFV), Multi-Access Edge Computing (MEC), and Cloud/Fog computing [4]. Network softwarization [10, 12] in 5G networks is intended to deliver future services and applications with greater agility and cost-effectiveness by employing software programming in the design, implementation, deployment, and management of network equipment/components/services [12, 13]. The E2E service QoE management and 5G network requirements (e.g. programmability, flexibility, and adaptability) are to be realized by network softwarization [4, 14]. Network softwarization and virtualization technologies are set to offer capabilities to developers and operators to build network-aware applications and application-aware networks. This would enable them to match their end-users' business demands in 5G and the beyond networks. However, new design and implementation is needed in different 5G network segments (e.g. RAN, transport networks, core networks, mobile-edge networks and network clouds) in order to achieve network softwarization goals. This is because each network segment needs a

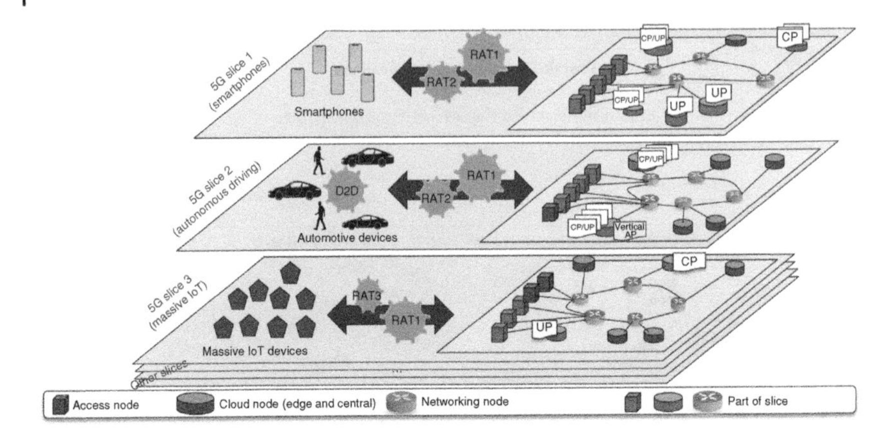

Figure 1.4 The NGMN 5G network slicing implementation.

different softwarization level and set of technical characteristics [15]. Figure 1.3 indicates software network technologies at different network segments in 5G systems.

For example, softwarization in mobile edge networks should be implemented using virtualized platforms based on SDN, NFV, and Information Centric Networking (ICN) [16, 17]. The design of most core networks and service plane functions in 5G networks have to be designed and implemented as Virtual Network Functions (VNFs) based on the envisaged SDN/NFV architectural principles. This approach would enable them to run on Virtual Machines (VMs), potentially over standard servers on Fog/Cloud Computing (CC) environments [18, 19]. That way, different network slices (smart cars) can use core networks and service VNFs based on the required latency and storage capacity of the requested service. In order to easily implement resource discovery and optimization mechanisms in the 5G control plane, the design of softwarized transport network can be done using appropriate interfaces in SDN/NFV infrastructures. This would allow various user applications and network-based services to be accommodated in 5G systems (Figure 1.4).

1.3 5G Service Quality and Business Requirements

5G networks promise to be more reliable on smart devices and facilitate ultra-fast video downloads. Different domains such as D2D/M2M, health (e.g. e-health, telemedicine), industrial 4.0, entertainment, intelligent transportation are to be facilitated with various 5G applications and services. Different requirements for

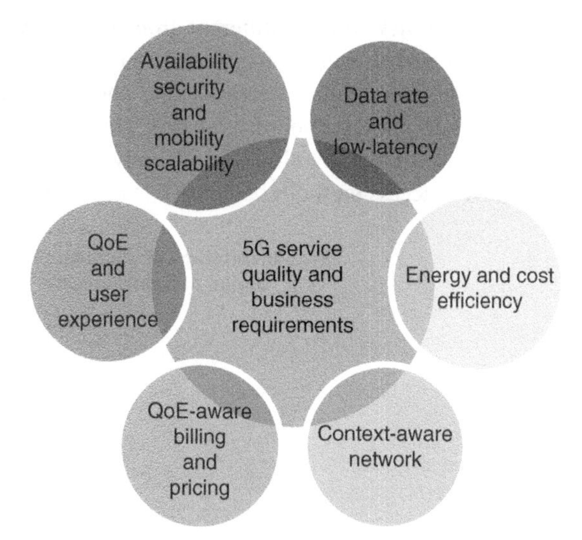

Figure 1.5 5G service quality and business requirements.

these 5G applications will be required in order to enhance their performance. For 5G to meet these performance requirements, new ways with intelligent network traffic management, caching, mobility and offload schemes, as well as enhanced capacity (e.g. small cells deployment) will have to be developed. Some of the 5G vision is to provide an increase of network bandwidth, coverage, and Internet connectivity, a massive reduction in energy consumption on smart devices and lower latency. With lower latency in 5G, downloading a large video file on a customer's device will be achieved within a few seconds. Some of the identified requirements of 5G networks as illustrated in Figure 1.5 that will impact the users experience for streaming services and voice calls include data rate and low latency, availability, security, mobility and scalability, QoE and user experience, energy, and cost efficiency.

1.3.1 High User Experienced Data Rate and Ultra-Low Latency

5G networks are expected to provide increased data rates which are 10 times (e.g. 1–10 Gbs) compared with the current 4G LTE networks that provides a theoretical data rate of 150 Mbps [4]. 5G connections will be able to support future applications (e.g. tactile Internet, mobile telepresence with 3D rendering capabilities, AR/VR and 3D gaming) that require a 100× increase in an achievable data rate. As such, unlimited broadband experiences and high QoE-oriented services will be guaranteed to the end-users even in crowded areas such as stadiums, trains, and shopping malls. 5G will also provide a 10× reduction in E2E latency. Lower latency offered by 5G mobile networks will enhance user experiences to do entirely new things, different from how they are done today. 5G lower latency will benefit multiplayer mobile gaming, factory robots, high-definition streaming from cloud-based technologies and enhanced VR devices (e.g. Google Glass), autonomous and self-driving cars. For example, premium user experience and YouTube Video downloads in 5G networks could be done within seconds regardless of access methods being used (Figure 1.6).

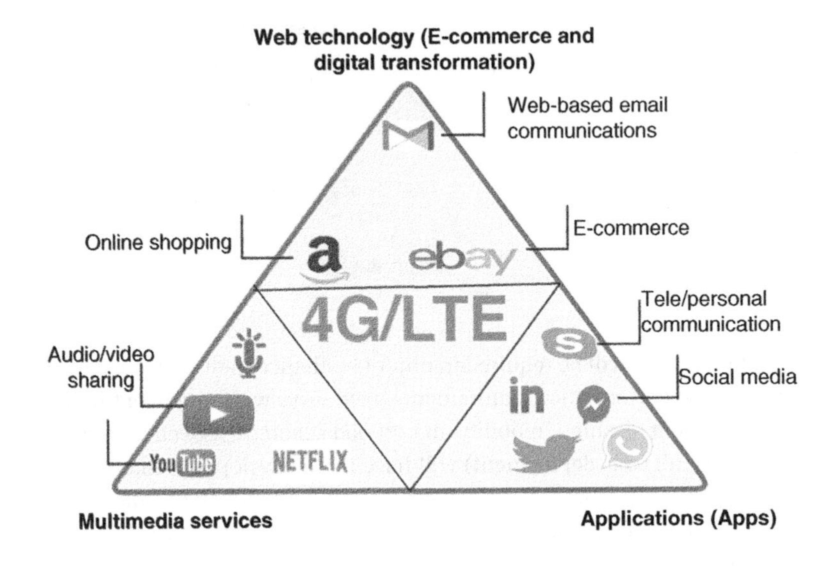

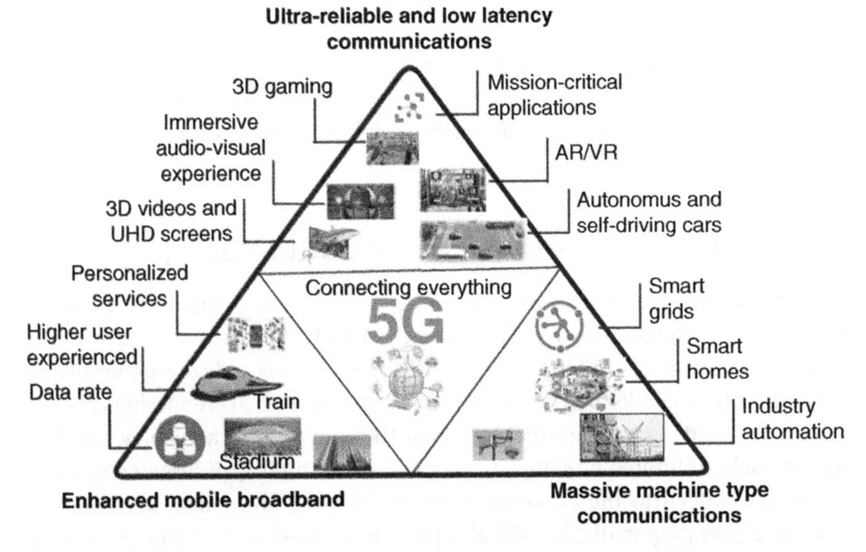

Figure 1.6 Current and future 5G network capabilities to support various use cases.

1.3.2 Transparency, Consistency, User's QoE Personalization, and Service Differentiation

5G networks will offer almost 100% consistent network coverage for "anytime, anywhere" connectivity. Users on 5G networks will get network coverage from any geographical locations. Fluctuations in the network performance, service disruptions, and unpredictable nature of wireless channels have to be kept at a minimal level for 5G to guarantee the required end-users' QoE provisioning. 5G should hide its architectural complexity and allow a high level of transparency. The 5G transparent network in turn would deliver the "best experience" during remote services and data/information delivery from cloud-service providers to their customers. Personalized QoE management strategies and service quality differentiation are required in 5G networks especially in Video-On-Demand/Live-TV services. To achieve this, specific mechanisms of charging regarding video quality levels have to be developed in 5G networks. These mechanisms should be able to learn from a user's video content consumption patterns in real-time or on-demand basis.

1.3.3 Enhanced Security, Mobility, and Services Availability

Enhanced security is required in 5G networks to support mission-critical applications (e.g. industrial control, smart grids, telemedicine, automotive, and public safety) that have stringent security requirements for defending against intrusions and uninterrupted operations. New trust business models with enhanced security as a service, identity management, and privacy protections in 5G network have to be supported. It is important to mention that 5G-based security architecture should be developed to support security attributes for different network slices. These security aspects and related challenges in 5G and beyond networks will be discussed in Chapters 9 and 10. To simplify security management in 5G systems, a uniform security management framework for multi-vendor environment has to be developed. That way, it will be easy to offer differentiated security to different services in 5G networks. Today, the density of users and the heterogeneity of smart devices that need an enhanced service availability are increasing at unprecedented speed. Efficient mobility management solutions that will facilitate seamless user mobility in 5G and beyond networks are of vital importance. Service continuity on each device without losing connectivity should be at the forefront in the 5G network design. As a consequence, handover procedures and a topology-aware gateway selection and relocation algorithms [20] in 5G networks need to be re-evaluated and/or re-designed. The Distributed Mobility Management (DMM) [21, 22], proposals for 5G can be a starting point in overcoming the current mobility management limitations.

1.3.4 Seamless User Experience, Longer Battery Life, and Context Aware Networking

5G networks should provide a consistent user experience regardless of the end-user's geographical location. The Key Quality Indicators (KQIs) are the achievable quality experienced by the end-user in terms of data rates and lower latency. 5G should have attributes that enable seamless integration of both inter-Radio Access Technology (RAT) and intra-RAT handovers mechanisms. That way, 5G networks can allow interruptions in the order of few milliseconds for services such as ultra-high definition video, 3D gaming, or the tactile Internet. 5G is anticipated to connect thousands of smart devices and IoT applications that involve a battery-operated sensor networks by 2023. Deploying sensor networks in 5G will be possible only if much longer battery life and low energy consumption will be guaranteed for several years on these smart devices. 5G should be a context-aware network that provides small-cell and macro-network as well as multi-RAT of different capabilities. It should also provide both device-awareness and application-awareness with associated user subscription context (e.g. operator preferences) and QoS/QoE-service requirements. That way, 5G should be able to adapt dynamically to the end-user's demands, application, and device's requirements.

1.3.5 Energy and Cost Efficiency, Network Scalability, and Flexibility

Cost efficient in 5G will be achieved by NFV [23]. NFV promise to offer better service agility and reduce the OPerational EXpenditure (OPEX) and CAPital EXpenditure (CAPEX) through lower-cost flexible network infrastructures in 5G. It is worth stressing that, NFV will decrease the deployment time of new services and applications to the market and support changing QoE-based business requirements from customers. While the current base stations (BSs) in 4G LTE networks contribute between 60% and 80% of the whole cellular network energy consumption [4], 5G network should cope with the increasing density of users and time variant traffic patterns. 5G should provide proper mechanisms that make the network more energy efficient by optimizing the number of active elements as the traffic grows/decreases. To avoid energy wastage in 5G networks, different network load variations can be managed using the deployment of low-cost, low-power access nodes such as small cells and wireless relays. The forefront that has been promoting the energy-efficient 5G networks include the Green 5G Mobile Networks (5GrEEn) [24], GreenTouch [25], Energy Aware Radio and neTwork tecHnology (EARTH) [26]. With flexibility, 5G network architecture should allow the RAN and the core network to evolve and scale independently of each other. That way, changes made in one layer should not affect another layer in the network. In addition, 5G network should be scalable enough and handle different signaling traffic from large number of IoT devices. SDN will be a key technology in supporting scalability in both the user plane and the control plane [4].

1.4 5G Services, Applications, and Use Cases

5G services are expected to cover a wide range of applications, which are categorized into three groups, namely, (i) enhanced Mobile Broadband (eMBB), (ii) Ultra-reliable and Low Latency Communications (URLLC), and (iii) massive Machine-Type Communications (mMTC) or massive Internet of Things (MIoT). The eMBB provides network connectivity with enhanced user experience for both indoors and outdoors. It also enables a wide network coverage in challenging conditions such as crowded areas (e.g. stadiums, trains) and shopping malls. Some of the sub-use cases include hot spots - broadband access in dense areas, broadband everywhere to provide a consistent user experience, public transport systems (high-speed trains), smart offices and specific events (e.g. stadiums, live concerts) where hundreds of thousands users have to be served with high data rate and low latency.

URLLC indicates mission-critical IoT applications ranging from smart grids to intelligent transportation systems, industrial Internet, autonomous and self-driving and remote surgery that require better reliability, improved energy efficiency, sub-millisecond latency, and massive connection density. URLLC signifies also different 5G applications that need an E2E security and almost 99.999% reliability in their performance. The URLLC sub-use cases include process automation, automated factories, tactile interaction (e.g. robotic controls in manufacturing, remote medical care and autonomous cars). In addition, it also includes emergency, disasters management and public safety, and urgent healthcare (e.g. remote surgery, diagnosis, and treatment).

5G networks should also support enhanced multimedia where high media-quality experience will be provided to the end users everywhere and meet their demands. The most significant driving factors of enhanced media use case in 5G systems include the recent developments of 4K/8K video resolutions, 3D videos, and 360° interactive videos. 5G will realize the vision of enhanced multimedia and offer even a seamless mobile TV experience to users using features such as enhanced broadcast/multicast, faster data rates, and higher data capacity. From the end user's perspective, the QoE will be the main performance metric in 5G networks. Delivering services with high QoE to the end users will be vital in supporting rich media services such as video gaming, 4K/8K/3D video that customers are accustomed to nowadays. 5G will support and deliver the multimedia services such as high-definition voice and videos (4K/8K) to the end-users.

Table 1.1 A Summary of user experience requirements in 5G networks.

Use case	Mobility (km/h)	E2E latency (ms)	User data rate (Mbps)
Mobile broadband in vehicles (e.g. cars, trains)	On demand, up to 500	10	DL: 50; UL: 25.
Broadband access in dense areas (e.g. urban city)	On demand, up to 100	10	DL: 300; UL: 50.
Broadband access in a crowd (e.g. stadiums, shopping malls)	Pedestrian	10	DL: 25; UL: 50.
Indoor ultra-high broadband access	Pedestrian	10	DL: 1000; UL: 500.

Connected vehicles will be an important driver for 5G networks. Furthermore, smart and autonomous vehicles represent a new context for experiencing networked communication and multimedia services. Smart and autonomous vehicles require fast and reliable communication networks such as 5G. In addition, smart vehicles can or have to be controlled remotely by an external operator. Remote control will also be enabled by 5G networks that can provide guaranteed network resources and enable very short response times which are crucial in applications such as telepresence and networked multimedia services with excellent QoE. The 3GPP defines four types of use cases including vehicle-to-infrastructure (V2I), vehicle-to-vehicle (V2V), vehicle-to-network (V2N,) and vehicle-to-pedestrian (V2P). Again, high reliability and mobility, ultra-low latency for warning signals, mitigating road accidents, and higher data rates for sharing sensor data and information are features that must be supported in 5G networks. Table 1.1 shows a summary of user experience requirements in 5G networks.

1.5 5G Standardization Activities

The development of 5G networks has been progressing at a fast pace from the industry and academia. 5G standardization activities have been taking place in different bodies (e.g. 3GPP, ITU-T IMT2020, ETSI, IETF), innumerable fora and alliances, consortia and collaborative projects (e.g. 5GPP, GSMA, NGMN, TMN, BBF, and OneM2M). 5G standardization work in 3GPP started in early 2015. The main 5G activities in 3GPP [27] include the development of network technologies for cellular telecommunications regarding the core network, RAN, transport network, and service capabilities. As part of 3GPP Release 14, the 3GPP Services and Requirements Working Group defined the high-level 5G service requirements. The 3GPP Release 15 specifies a "standalone" 5G network with a new radio system. Release 16 specifies topics related to LAN support in 5G, Vehicle-to-everything (V2X) application layer services, wireless and wireline convergence for 5G, multimedia priority Service and 5G satellite access. The 3GPP Release 17 has been finalized in September 2021 covering studies on management aspects of MEC, protocol enhancement for next-generation real time communication beyond 5G systems [28]. In addition, it also studies the multi-tenancy, multi-domain concept and 5G network slicing management. The 3GPP Technical Specification Groups (TSG) are divided into three categories focusing on (i) 5G RAN, (ii) 5G Core Networks and Terminals (CT), and (iii) 5G Services and Systems Aspects (SA). Table 1.2 shows a summary of standardization efforts from academia and industry.

ITU-T IMT2020 [29] Study Group (SG) 13 has been working on 5G network softwarization including network slicing and orchestration, integrated network management, overall 5G architecture and key requirements, and mobility management framework over reconfigurable 5G networks. ITU-TSG13 has also introduced a new focus group to study on the application of machine learning

Table 1.2 A summary of standardization efforts from the academia and industry.

Name and working group		Description of 5G-related work
ITU-T IMT-2020	SG13	Network requirements and functional architecture, 5G network softwarization and slicing, 5G network management and orchestration requirements
3GPP	TSG RAN,TSG SA, TSG CT	LAN support in 5G, Vehicle-to-everything (V2X) services, multimedia priority Service, 5G network slicing management [27]
IEEE	802.11 WG	Developing use cases, framework/architectures, requirements and standards for security, performance, reliability and bootstrapping procedures for SDN and NFV
IETF/IRTF	SDN RG	Defines key specifications and requirements for virtualized functions that evolve IP protocols for supporting the 5G network virtualization. Other 5G activities include SFC to enable the dynamic creation and linkage of VNFs such as packet data gateway, serving gateway into a single path in 5G systems
ETSI	—	To develop the 5G network architecture that will enable data connectivity and services for 5G system

Note: WG = Working Group; CT = Core Network & Terminals; SG = Study Group; SA = Service & Systems Aspects.

and artificial intelligence for future 5G networks [30]. The IETF in the context of 5G networks specifies data models, service chaining, QUIC and user plane protocol, and architectural analysis on 3GPP 5G system. The IETF [31] also defines key specifications and requirements for virtualization functions that evolve IP protocols to support 5G network virtualization. The IETF is the pioneer of a Service Function Chaining (SFC) to enable the dynamic creation and linkage of VNFs such as packet data gateway, serving gateway into a single path in 5G systems. Moreover, the IETF works on protocols for distributed networking, path computation, and segment routing in order to meet the 5G NR. ETSI is actively working in collaboration with the 3GPP on developing the system architecture to provide data connectivity and services for 5G system. ETSI has also been working on security (authentication) issues in 5G, QoS/QoE, policy control, and charging. The works in IEEE on 5G have been focusing on developing specification for physical and Medium Access Control (MAC) layer to provide wireless connectivity for fixed, portable and moving stations within a local area in 5G systems [5]. The IEEE is also working on specifying the 5G application scenarios of the "Tactile Internet" and developing its architecture. The main goal is to promote and facilitate the rapid realization of the Tactile Internet in 5G systems and beyond. The NGMN [32] 5G initiatives include defining requirements and architecture for 5G, network management and orchestration, and 5G security.

1.6 Conclusion

This chapter presents an introduction to 5G networks. It provides the history of mobile communication systems followed by the vision and motivation of 5G systems. The 5G service quality and business requirements as well as future services, applications and use cases are presented. The chapter further presents standardization activities of 5G networks from different bodies, consortia, and alliances in industry and academia. It is evident that some of the requirements that will enhance the delivery of multimedia service quality include: high user experienced data rate, ultra-low latency, transparency, consistency, user's QoE personalization and service differentiation, enhanced security, mobility and services availability, seamless user experience, longer battery life and context aware networking, energy and cost efficiency, network scalability, and flexibility.

Bibliography

1 Agiwal, M., Roy, A., and Saxena, N. (2016). Next generation 5G wireless networks: a comprehensive survey. *IEEE Communication Surveys and Tutorials* 18 (3): 1617–1655.

2 Osseiran, A., Boccardi, F., Braun, V. et al. (2014). Scenarios for 5G mobile and wireless communications: the vision of the METIS project. *IEEE Communications Magazine* 52 (5): 26–35.

3 Andrews, J.G., Buzzi, S., Choi, W. et al. (2014). What will 5G be?. *IEEE Journal on Selected Areas in Communications* 32 (6): 1065–1082.

4 Barakabitze, A.A., Barman, N., Ahmad, A. et al. (2019). QoE management of multimedia services in future networks: a tutorial and survey. *IEEE Communication Surveys and Tutorials* 22 (1): 526–565.

5 Barakabitze, A.A., Ahmad, A., Mijumbi, R., and Hines, A. (2020). 5G network slicing using SDN and NFV: a survey of taxonomy, architectures and future challenges. *Computer Networks* 167: 1–40.

6 Chen, S. and Zhao, J. (2014). The requirements, challenges, and technologies for 5G of terrestrial mobile telecommunication. *IEEE Communications Magazine* 52 (5): 36–43.

7 Ali, M.A. and Barakabitze, A.A. (2015). Evolution of LTE and related technologies towards IMT-advanced. *International Journal of Advanced Research in Computer Science and Software Engineering* 5 (1): 16–22.

8 Varela, M., Zwickl, P., Reichl, P. et al. (2015). From service level agreements (SLA) to experience level agreements (ELA): the challenges of selling QoE to the user. *Proceedings of IEEE ICC QoE-FI*, June 2015.

9 Yousaf, F.Z., Bredel, M., Schaller, S., and Schneider, F. (2018). NFV and SDN-key technology enablers for 5G networks. *IEEE Journal on Selected Areas in Communications* 35 (11): 2468–2478.

10 Afolabi, I., Taleb, T., Samdanis, K. et al. (2018). Network slicing and softwarization: a survey on principles, enabling technologies and solutions. *IEEE Surveys and Tutorials* 20 (3): 2429–2453.

11 Barakabitze, A.A., Sun, L., Mkwawa, I.-H., and Ifeachor, E. (2016). A novel QoE-aware SDN-enabled, NFV-based management architecture for future multimedia applications on 5G systems. *2016 8th International Conference on Quality of Multimedia Experience (QoMEX)*, Lisbon, Portugal.

12 Condoluci, M., Sardis, F., and Mahmoodi, T. (2016). Softwarization and virtualization in 5G networks for smart cities. *International Internet of Things Summit* 169 179–186.

13 Afolabi, I., Taleb, T., Samdanis, K. et al. (2018). Network slicing and softwarization: a survey on principles, enabling technologies and solutions. *IEEE Communication Surveys and Tutorials* 20 (3): 1–24.

14 Barakabitze, A.A., Mkwawa, I.-H., Sun, L., and Ifeachor, E. (2018). QualitySDN: Improving video quality using MPTCP and segment routing in SDN/NFV. *IEEE Conference on Network Softwarization*, May 2018.

15 Sanchezl, J., Yahia, I.G.B., Crespi, N. et al. (2014). Softwarized 5G networks resiliency with self-healing. *2014 1st International Conference on 5G for Ubiquitous Connectivity (5GU)*, pp. 229–233, November 2014.

16 Ravindran, R., Chakraborti, A., Amin, S.O. et al. (2017). 5G-ICN: Delivering ICN services over 5G using network slicing. *IEEE Communications Magazine* 55 (5): 101–107.

17 Barakabitze, A.A. and Xiaoheng, T. (2014). Caching and data routing in information centric networking (ICN): the future internet perspective. *International Journal of Advanced Research in Computer Science and Software Engineering* 4 (11): 26–36.

18 Liu, Y., Fieldsend, J.E., and Min, G. (2017). A framework of fog computing: architecture, challenges, and optimization. *IEEE Access: Special Section on Cyber-Physical Social Computing and Networking* 5: 25445–25454.

19 Yi, S., Hao, Z., Qin, Z., and Li, Q. (2015). Fog computing: platform and applications. *Proceedings of the 3rd IEEE Workshop on Hot Topics in Web Systems and Technologies (HotWeb)*, pp. 73–78, November 2015.

20 Heinonen, J., Korja, P., Partti, T. et al. (2016). Mobility management enhancements for 5G low latency services. *IEEE ICC2016-Workshops: W01-3rd Workshop on 5G Architecture (5GArch 2016)*, May 2016.

21 Giust, F., Cominardi, L., and Bernardos, C.J. (2015). Distributed mobility management for future 5G networks: overview and analysis of existing approaches. *IEEE Communications Magazine* 53 (1): 142–149.

22 Nguyen, T.-T., Bonnet, C., and Harri, J. (2016). SDN-based distributed mobility management for 5G networks. *IEEE Wireless Communications and Networking Conference(WCNC)*, April 2016.

23 Mijumbi, R., Serrat, J., Gorricho, J.L. et al. (2016). Management and orchestration challenges in network functions virtualization. *IEEE Communications Magazine* 54 (1): 98–105.

24 Olsson, M., Cavdar, C., Frenger, P. et al. (2013). 5GrEEn: Towards green 5G mobile networks. *Proceedings of the 9th International Conference on Wireless and Mobile Computing, Networking and Communications (WiMob)*, October 2013.

25 FI-Core (2016). The GreenTouch Project. http://www.greentouch. (accessed 06 March 2018).

26 Gruber, M., Blume, O., Ferling, D. et al. (2009). EARTH – Energy aware radio and network technologies. *IEEE 20th International Symposium on Personal, Indoor and Mobile Radio Communications*, October 2009.

27 Bertenyi, B. (2019). 3GPP system standards heading into the 5G era, November 2019.

28 3rd Generation Partnership Project (3GPP). The 3GPP Release 17, November 2019.

29 Hans (Hyungsoo), K.I.M. (2019). IMT2020/5G Standardization in ITU-T Study Group 13, March 2019.

30 Andreev, D. (2019). Overview of ITU-T activities on 5G/IMT-2020, March 2019.

31 Homma, S., Miyasaka, T., and Voyer, D. (2019). Overview of ITU-T activities on 5G/IMT-2020, March 2019.

32 NGMN Alliance (2015). 5G White Paper, February 2015.

2

5G Network Management for Big Data Streaming using Machine Learning

Multimedia big data have been rapidly captured, analyzed, and disseminated as a result of the rapid development of digital multimedia sensing and communication streaming and networking technologies [1]. Multimedia big data have become an integral part of today's big data because the majority of Internet video traffic today is delivered via HTTP-based adaptive streaming (HAS) at the end-user's device. Such large amounts of multimedia data present both benefits and challenges in terms of managing, searching, and comprehending them, necessitating an understanding of how humans perceive cognitive concepts and the end-user's QoE [2]. We can now considerably increase the intelligence of multimedia analysis thanks to recent breakthroughs in machine learning (ML) techniques. ML for massive multimedia analytics is becoming a hot topic in multimedia and computer vision research. This has resulted in a significant increase in research efforts devoted to solving problems in the field of multimedia computing. In recent years, ML technology has been widely applied to solve complicated and difficult real-world image processing challenges, such as visual surveillance, smart cities, and social multimedia networks. This chapter provides background information regarding the application of ML and artificial intelligence (AI) in managing big data streaming in future softwarized and virtualized 5G networks [3]. It will also provide the concepts for optimization of data management with ML in softwarized 5G networks. The chapter further presents state-of-the-art research work on multimedia big data analytics by providing the current big data 5G frameworks and their applications in multimedia analyses [4].

2.1 Machine Learning for Multimedia Networking

Due to the plethora of new multimedia applications and services that have emerged in the last decade, a massive amount of multimedia data has been generated for the purposes of advanced multimedia research [5]. ML has been

Multimedia Streaming in SDN/NFV and 5G Networks: Machine Learning for Managing Big Data Streaming, First Edition. Alcardo Barakabitze and Andrew Hines.

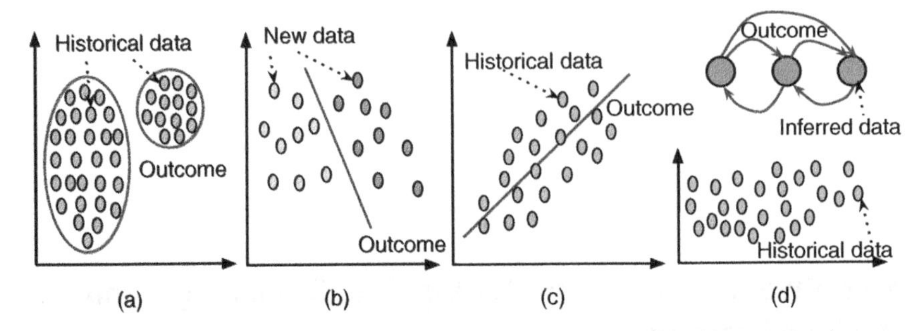

Figure 2.1 Problem categories that benefit from ML. Letters a, b, c, and d indicates clustering, classification, regression, and rule extraction, respectively.

enjoying a meteoric rise in applications that solve issues and automate processes across a wide range of industrial domains including multimedia streaming. This is mostly due to the explosion in data availability, considerable advances in ML techniques, and advancements in computing capability [6]. Without a doubt, ML techniques have been used to solve a variety of simple and complicated challenges in network operation and management by relying heavily on big data. ML focuses on extracting information from massive volumes of data, detecting and identifying underlying patterns using a variety of statistical measurements to increase its ability to understand new data and providing more effective outcomes. Clearly, some factors should be fine-tuned at the outset to improve and solve problems that have a large representative dataset. As indicated in Figure 2.1, ML techniques are designed specifically for identifying and exploiting hidden patterns in big data. That way, ML can (a) describe the outcome as a grouping of big data for clustering problems, (b) predict the outcome of future events for classification and regression problems, and (c) evaluate the outcome of a sequence of data points for rule extraction problems [6].

A generic approach and the process flow used for building ML-based solutions are illustrated in Figure 2.2. Key components of this process flow include (a) Data Collection, (b) Feature Engineering, (c) Model Optimization and Training and ground truth of Data, (d) testing, and (e) Model Validation and Evaluation.

2.1.1 Data Collection

The data collection is the first step for building ML-based solutions where a large dataset for a given problem is collected either offline or online. An online dataset is collected in real-time during multimedia streaming. The data collected in this phase can be used as input for retraining the ML-based model or providing feedback to the ML model. Offline data is collected to gather historical data which can be employed in the training and testing phase of the ML-model. Offline data can

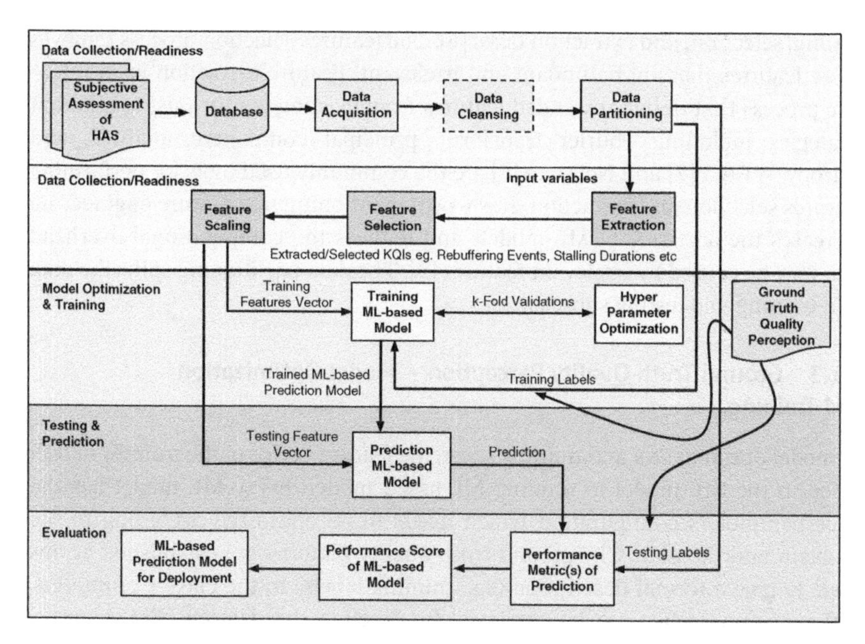

Figure 2.2 Process flow diagram for ML solution in multimedia streaming service.

be collected using different repositories such as the Measurement and Analysis on the WIDE Internet (MAWI) Working Group Traffic Achieve, the Waikato Internet Traffic Storage (WITS), and UCI Knowledge Discovery in Databases Achieve [6]. To provide a good control for various aspects regarding, for example data sampling rate, location (core network or edge network) and monitoring duration, the data collection process for both offline and online, and the monitoring (e.g. active or passive) and measurement tools can be used. The measurement network traffic, for example the probe packets are injected in the network in the case of active monitoring, whereas data is collected based on the observation of the actual network traffic.

2.1.2 Feature Engineering

It is important to note that the collected data in the first phase may be incomplete or consist noisy. Therefore, the data have to be cleaned before using it for learning. Before training the model, important features for learning and inference have to be extracted from the collected data. For example in multimedia networking, packet-level features (e.g. statistics of packet size) are extracted or driven from the packets during a multimedia transmission. Feature engineering is the most important part of developing a ML-based solution and it includes the feature scaling, selection, and extraction tasks [6]. The features selection process removes those features that are redundant and irrelevant. Feature extraction is an intensive process that derives extended features from existing features using different strategies including Fourier transform, principal component analysis, and entropy. WEKA [7] and NetMate [8] are the commonly used tools for performing features selection and extraction. It is worth mentioning that feature engineering increases the accuracy of ML models and reduces the computational overhead that may be caused by irrelevant features [9]. The data partitioning splits the data into training and testing subsets.

2.1.3 Ground Truth Quality Perception – Model Optimization and Training

In model optimization and training stage, the feature vector of the training subset is fed to the ML model to training ML-based model. Every ML model has the hyperparameters configuration which needs to be optimized according to the problem and the data. The ground truth quality patterns are established at this stage to give a formal description (e.g. training labels) to the classes of interest. Different approaches can be employed for labeling the datasets. For example, the application signature pattern-matching techniques [10] can be used to establish ground truth for classes in the training dataset during multimedia traffic classification. Other labeling techniques include unsupervised ML techniques, for example Auto-Class using EM [11]. It is worth noting that the ground truth (model optimization and training) derives the accuracy from the ML models. However, the size of the training data of once class of interest can greatly affect the performance of the ML model.

2.1.4 Performance Metrics, Testing, and Model Validation

The last phase is the testing and ML model evaluation which is done by comparison with testing labels based on the considered performance metrics. The performance metrics can be employed to perform measurement of different aspects of the ML-model such as robustness, reliability, complexity, and its accuracy. Once the trained ML-based model meets the performance criteria, then it can be deployed for QoE-aware network and service management.

2.2 Machine Learning Paradigms

ML is the foothold of AI. The learning paradigms in ML that influence the data collection, feature engineering, and the establishment of ground truth (training and optimization) are divided into four categories, namely: supervised, unsupervised, semisupervised, and reinforcement learning. The training data are basically the

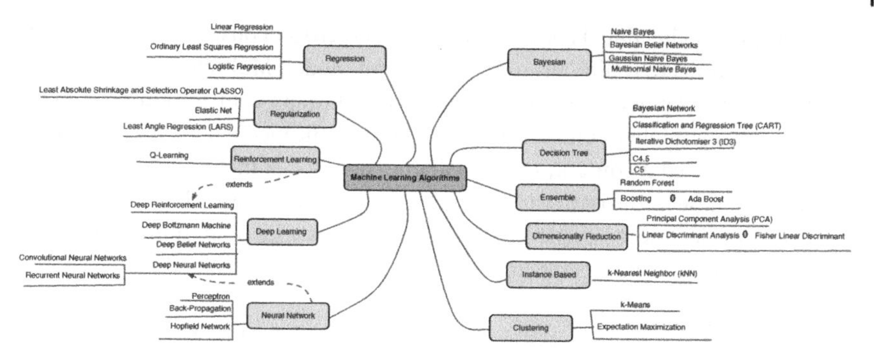

Figure 2.3 Classification of ML algorithms.

dataset that is used to construct the ML model. The training labels are associated with the training data as long as the user understands well the description of the data [12]. Given the datasets, the objective of ML is to infer the outcome that is perceived as the identification of the membership to a class of interest [6]. Supervised learning employs labeled training datasets in creating ML-based models. This ML technique identifies different patterns or behaviors while the training datasets are known. Regression and traffic classification problems are normally solved by supervised learning methods where patterns of discrete or continuous valued outcomes are predicted. On the other hand, unsupervised learning employs unlabeled dataset for creating ML-based models. Unsupervised learning is suitable for solving clustering problems where patterns are discriminated in the data. In multimedia networking, density estimation and outliers detection issues can pertain to grouping various instances of attacks based on their similarities.

Reinforcement learning (RL) is an agent-based iterative process that is used for modeling decision-making challenges. An agent in RL continuously interacts and explores the environment to identify the best action. The agent learns to maximize the expected outcome through constant exploration and exploitation of knowledge. The training dataset in RL consists of a set of state-action pairs. The actions are rewarded or penalized. To optimize the cumulative reward, the agent uses the feedback from the environment to learn the best sequence of actions or policy. Because of its capability for making cognitive choices including planning and scheduling, RL has been used for resource management over wireless networks, user scheduling, and content caching for mobile edge networks (Figure 2.3).

2.3 Multimedia Big Data Streaming

Multimedia big data [13, 14] is defined as a large-scale signals which consists of unknown complex structures, involving time statistics, motion trajectory, spatial statistics, human factors, and interview correlations with structured singularities

[15]. Multimedia big data is often multimodal, heterogeneous, and unstructured, attributes that make it difficult to represent and model. Multimedia big data are a type of datasets where data are heterogeneous, human-centric, and have more media types and higher volume compared to nonmultimedia big data [15]. From media type to media content, multimedia big data focus more on humans than machines. Datasets consist of various types of video data. This includes interactive and camera video, stereoscopic 3D video, 3D virtual worlds, and immersive video. This makes multimedia big data to have higher level of complexity than typical big data such as text-based data.

Multimedia big data come with some issues and challenges as well as opportunities for multimedia applications/services especially while heading toward 6G networks. Multimedia big data can be generated and acquired from various sources such as cloud gaming, ubiquitous portable mobile devices (smartphone, etc.), multimedia sensor, video lectures, immersive videos, the Internet of Things (IoT) [16], and social media video [15]. This causes the modeling and the overall representation of multimedia big data to be very challenging because data originates from different sources (cyber space, social, and physical). The life cycle of multimedia big data is indicated in Figure 2.4 consisting of acquisition, compressing, storage, processing, understanding, assessment, computing, and security. The acquisition stage entails the generation of multimedia big data from different heterogeneous sources including camera video, smartphone devices, social media videos, etc. The generated raw data have to be compressed by using signal processing and transformation techniques. The compressed raw data have to be stored in a repository. Multimedia big data with large-scale high-dimensional data with a tolerable processing time needs to be processed rapidly and continuously subject to storage and time constraints. The assessment of multimedia big data is very important to

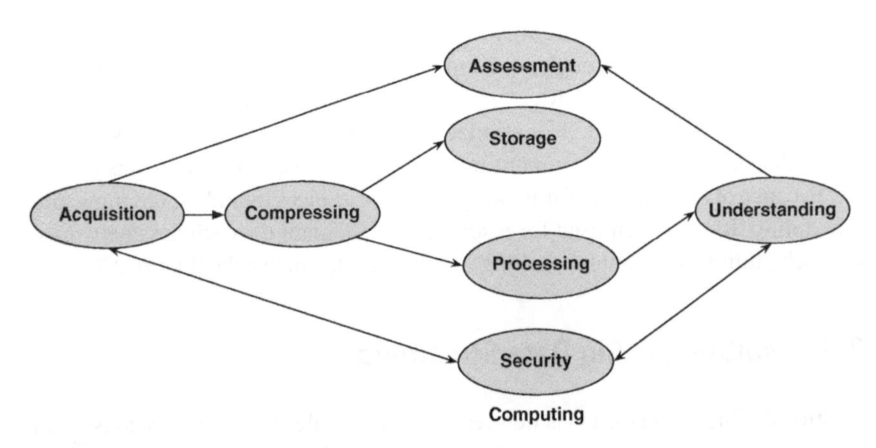

Figure 2.4 Multimedia big data.

monitor and analyze network traffic for improving the user's QoE and optimize network resource allocation [2].

Multimedia big data have various formats and different distribution attributes which also have different size, volume, and even types. They cannot be analyzed in the same way because of different patterns. Unstructured data are rapidly growing in both quality and quantity. Therefore, it is very important to design new ML-based models that can provide analytics of complex and heterogeneous data spaces including multimedia big data [1].

2.4 Deep Learning for IoT Big Data and Streaming Analytics

A massive number of sensing devices collect and/or generate varied sensory data throughout time for a variety of industries and applications in the IoT era [17]. These devices will provide large or quick/real-time data streams, depending on the application. The goal of the Internet of Things (IoT) is to turn everyday objects into intelligent ones by utilizing a variety of cutting-edge technologies, such as Internet protocols, data analytics, embedded devices, and communication technologies [18]. Numerous business prospects and applications such as multimedia streaming are anticipated to result from the potential economic impact of IoT, which is also anticipated to hasten the economic growth of IoT-based services. According to the IoT global economic effect research [19], the annual economic impact of IoT in 2025 would be between 2.7 and 6.2 trillion.

IoT data exhibit the following characteristics including (i) large-scale streaming data, (ii) heterogeneity, and (iii) time and space correlation. For IoT applications, a wide variety of data capturing devices are dispersed and deployed, and they continuously provide data streams. A significant amount of continuous data results from a large-scale streaming data. Data heterogeneity results from the collection of different information by various IoT data capture devices [20]. In addition to big data analytics, IoT data necessitates the development of a brand-new type of analytics called fast and streaming data analytics to support applications such as multimedia streaming services that require time-sensitive in high-speed data streams and real-time or near real-time) operations. IoT data can be streamed continuously or accumulated as a source of big data.

Figure 2.5 indicates three layers in the IoT big data streaming architecture consisting of IoT devices, edge devices or fog computing, and the IoT cloud computing platforms. The IoT devices and edge computing are where the hard and soft real-time analytics is performed. The deep learning models for IoT big data analytics can be performed at the IoT cloud layer on high-performance computing systems or cloud platforms. Data parallelism and incremental processing

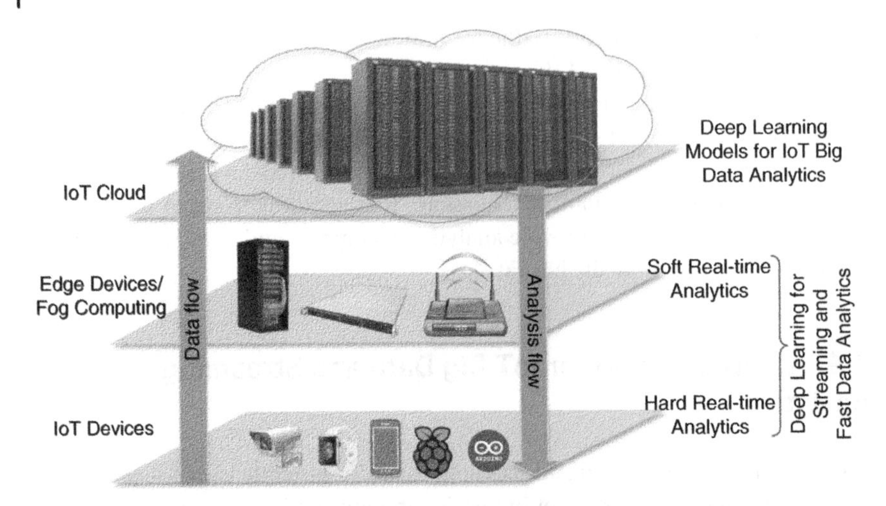

Figure 2.5 Deep learning models and IoT data creation at various levels to handle knowledge abstraction. Source: Mohammadi et al. [20] / with permission of IEEE.

are the foundations for the streaming data analytics on these frameworks. Data parallelism divides a huge dataset into numerous smaller datasets so that parallel analytics can be carried out on them all at once. When processing data incrementally, a pipeline of compute activities is used to quickly process a small batch of data. These methods simplify the time it takes the streaming data analytic framework to respond, but they are not the ideal answer for time-sensitive IoT applications. Fast analytics on IoT devices, however, come with their own set of difficulties, such as limited computation, storage, and power resources at the data source [20].

2.5 Intelligent QoE-based Big Data Strategies and Multimedia Streaming in Future Softwarized Networks

As of today, a tremendous amount of data are being generated from different sources such as the IoT, social networking websites (Facebook, Twitter, and Flicker), which is bound to increase even more in the coming years [20]. It is important to note that at this pace, the current static measurements of network and application performance will not be capable of keeping up to the changing dynamic landscape of future network softwarization. Therefore, creating a QoE-based dynamic model to correlate the resulting big data, probably, requires ML and AI that will move from the traditional lab-based modeling toward a QoE-driven live network predictive data analytics which is required for self-optimization and self-healing in future networks [21].

As stated by Cui et al. [22], on the one hand, software defined networking (SDN) can solve many issues of big data applications (e.g. big data acquisition, processing, transmission, and delivery in cloud data centers). On the other hand, as an essential network application, big data will have a profound impact on the overall operation and design of future SDN-based networks [23]. For example Wang et al. [24] introduce a cross-layer modular structure for big data applications based on SDN. Specifically, the run-time network configuration for big data applications is studied to jointly optimize the network utilization and application performance. Monga et al. [25] introduce SDN to big scientific data architectural models while an approach to big data analysis in SDN/Network functions virtualization (NFV)-based 5G networks is introduced by Barona López et al. [4]. However, while it will be challenging to meet QoS and QoE requirements and orchestrate virtual network functions (VNFs) without big data analytics, we note that the relationship between SDN, NFV, and big data is not yet studied, especially in the perspective of future networks. Therefore, it is vital to investigate new strategies for assigning and managing resources (e.g. in cloud data centers) to meet the Service-Level Agreements (SLAs)/experience level agreements (ELAs) of various big data applications in future networks [2].

2.6 Optimization of Data Management with ML in Softwarized 5G Networks

Multimedia big data consist of valuable information regarding content characteristics, users' behavior, and network dynamics. This information drives the system design and optimization of multimedia big data. The fundamental challenge is how to mine multimedia big data intelligence and further incorporate it into wireless multimedia networked systems.

2.6.1 A Multimodal Learning Framework for Video QoE Prediction

To solve these challenges, an architecture for content-aware video streaming using deep RL [26]. The proposed architecture consists of an agent, the deepQoE, and the streaming environment. The deepQoE component performs QoE modeling and interests analysis. The 3D ConvNets learns spatiotemporal video features after extracting 16 images for each video segment downloaded. The rectifier activation function video features are then fed into fully connected layers. The agent serves as a core for bitrate adjustments and the streaming environments. During video streaming sessions, the video server provides a set of video segments with equal

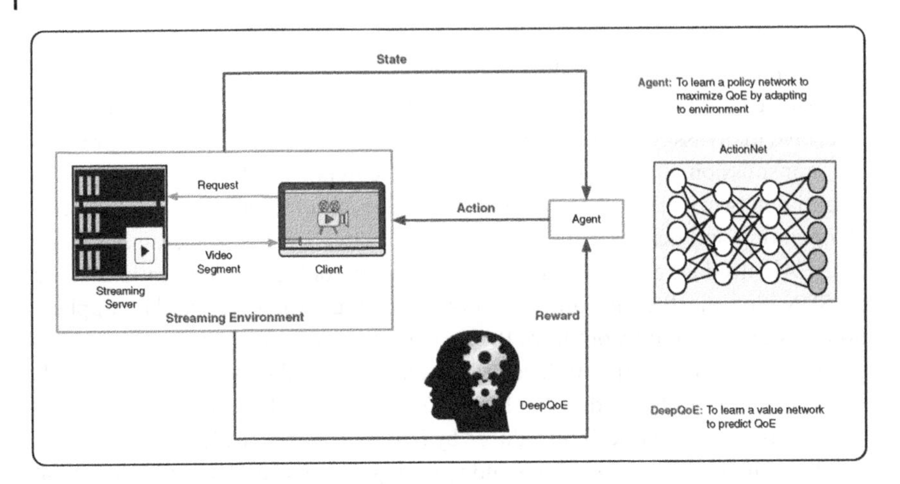

Figure 2.6 ML algorithms.

streaming duration and encodes each into different bitrate such as 240p, 360p, 480p, and 720p. During a video streaming session, the client MPD of the video file is requested, and the video bitrates are analyzed as well as the semantic information of the video content. The video player selects dynamically the video bitrate to download based on the measurement of the network parameters. It is important to mention that the RL problem is formulated as an interest-aware video bitrate adaptation mechanism where the optimal policy is learned online. Different strategies also regarding buffer occupancy and bandwidth allocation for streaming clients are described and presented. The control action module selects the appropriate video bitrate for the next video segment to be downloaded (Figure 2.6).

2.6.2 Supervised-Learning-based QoE Prediction of Video Streaming in Future Networks

A reference architecture for QoE prediction/measurement and management in future networks based on ML is proposed by authors in [27]. The ML-driven architecture shown in Figure 2.7 consists of various components in the end -to end (E2E) multimedia networking including OTT service providers, internet service provider (ISPs)/mobile network operator (MNOs), and the client side. The collaborative QoE-aware network and service management are performed by the OTT and ISPs/MNOs where the key quality indicators (KQIs) for measuring and estimating the end-user's QoE is accomplished by the OTT service providers.

2.6.3 OTT Service Providers

On the OTT side, the streaming media provides the required video content services to the clients through the Internet. The KQIs for QoE such as buffering events, duration, quality streaming switching rate and video bitrates are stored in the database.

2.6.4 ISP/MNOs

The reference architecture at the ISP/MNOs consists of the Data Plane (representing the ISP access network and core network), Control Plane (consisting of controller and NFV-MANO), and Application Plane. The NFV-MANO enables the deployment of network functions and dynamic provisioning of QoE-driven video services to the users. The SDN controller offers softwarization and programmability of network configurations and resources as well as QoE-optimization for improving the network performance [27]. The application plane consists of the following components: feature monitoring, feature engineering, predicted QoE, QoE-aware network resource management, and network control operations and trained ML model. The feature monitoring component monitors the KQIs for QoE in real time, whereas the feature engineering component select features and provide them to the trained ML model for QoE prediction based on the QoE feature vector. The trained ML model reports the estimated QoE to the QoE prediction module that stores the predicted/measured QoE data. This information becomes available to the QoE-aware network resource management component that optimizes the network resource based on the demands from the end-users.

2.6.5 Information Flow

The information flow and deployment options in the architecture are indicated using numbers (1–15). Steps 1–4 indicates the client requesting a video content from the media server using HTTP and the client adaptation algorithm adapting the video bitrates for the requested video segments. The adaptation is done by considering the feedback from HAS players and the current state of the network. The HAS players keep on downloading video segments from the media server based on the feedback received from the client adaptation algorithm. Step 5 shows that the clients keeps requesting video segments from the media server based on the feedback received from the client adaptation algorithm. The KQIs for the end-users' QoE from the HAS player are collected from the user-end probe using Steps 6–8. This information (KQI/input data) is then transferred and stored in the database on the OTT side.

In steps 9–11, the QoE prediction/measurements module performs the following functions: (i) retrieving QoE KQIs from the feature monitoring module,

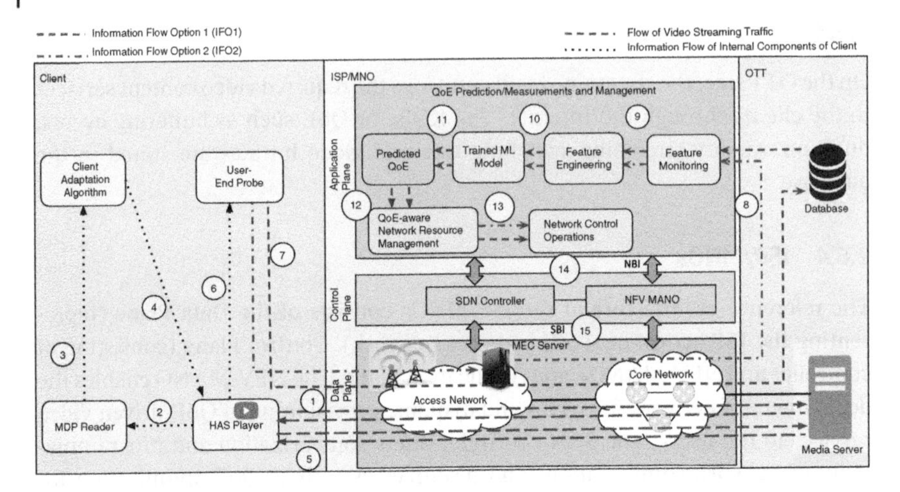

Figure 2.7 Architectural components, implementation, and deployment options using ML.

(ii) applying feature engineering to the input data to obtain the relevant feature vector, (iii) predicting QoE using a trained ML model based on the feature vector obtained from the feature engineering module, and (iv) storing predicted QoE for use by other network management applications [27]. Steps 12–13: The QoE-aware network resource management module optimizes and manages available network resources based on network management policies and SLA, using the expected QoE as an objective function (SLA). Steps 14–15: For QoE-aware network resource allocation, the network control operations module executes network operations such as network slicing in softwarized/virtualized networks, traffic flows engineering, and network services configuration [27]. It is worth mentioning that the reference architecture is able to provide QoE-aware network resource management by exploiting predicted QoE while enabling service and network management operations such as scaling of VNF, E2E network slicing, service prioritization, network traffic management, and radio resource allocation via SDN and NFV infrastructures.

2.7 Conclusion

The amount of multimedia data available every day is enormous and is growing at an exponential rate, creating a great need for new and more efficient approaches to large-scale multimedia streaming based on ML. Multimedia data have dramatically expanded as a result of the success of media streaming apps via the Internet in recent decades, putting increasing strain on the application provider's infrastructure, such as communication networking and video streaming architectures. This chapter presents ML for multimedia streaming and networking. This chapter provides a generic approach and the process flow used for building ML-based solutions. Key components of the ML-based process flow include (i) data collection, (ii) feature engineering, (iii) model optimization and training and ground truth of data, (iv) testing, and (v) model validation and evaluation. This chapter presents optimization of multimedia big data management architectures with ML in softwarized 5G networks. This includes a supervised-learning-based QoE prediction of video streaming in future networks and a multimodal learning framework for video QoE prediction.

Bibliography

1 Pouyanfar, S., Yang, Y., Chen, S.-C. et al. (2019). Multimedia big data analytics: a survey. *ACM Computing Surveys* 5 (1): 31580–31598. https://doi.org/10.1145/3150226.

2 Barakabitze, A.A., Barman, N., Ahmad, A. et al. (2020). QoE management of multimedia services in future networks: a tutorial and survey. *IEEE Communication Surveys and Tutorials* 22 (1): 526–565.

3 Barakabitze, A.A., Arslan, A., Rashid, M., and Hines, A. (2020). 5G network slicing using SDN and NFV: a survey of taxonomy, architectures and future challenges. *Computer Networks* 167: 1–40.

4 Barona López, L.I., Maestre Vidal, J., and García Villalba, L.J. (2017). An approach to data analysis in 5G networks. *Entropy* 19 (2): 1–23.

5 Wang, X., Zhu, W., Tian, Y., and Gao, W. (2020). Differential flow space allocation scheme in SDN based fog computing for IoT applications. *Proceedings of the 28th ACM International Conference on Multimedia*, pp. 4775–4776, October 2020.

6 Boutaba, R., Salahuddin1, M.A., Limam, N. et al. (2018). A comprehensive survey on machine learning for networking: evolution, applications and research opportunities. *Journal of Internet Services and Applications* 9 (16): 1–99.

7 Machine Learning Group (2017). University of Waikato. Weka. http://www.cs.waikato.ac.nz/ml/weka/ (accessed 11 August 2022).

8 Arndt, D. (2016). How to: calculating flow statistics using NetMate. netmate-flowcalc–calculating flow statistics from network traffic–Analytics library (openweaver.com).

9 Moore, A.W. and Zuev, D. (2005). Internet traffic classification using Bayesian analysis techniques. *CM SIGMETRICS Performance Evaluation Review*, ACM, pp. 50–60, December 2005.

10 Este, A., Gringoli, F., and Salgarelli, L. (2009). Support vector machines for TCP traffic classification. *Computer Networks* 53 (14): 2476–2490.

11 Erman, J., Mahanti, A., and Arlitt, M. (2006b). A geometric framework for unsupervised anomaly detection: detecting intrusions in unlabeled data. *Global Telecommunications Conference* 6: 1–6.

12 Huang, T., Zhang, R.-X., Zhou, C., and Sun, L. (2018). QARC: Video quality aware rate control for real-time video streaming based on deep reinforcement learning. *Proceedings of the 26th ACM International Conference on Multimedia*, October 2018.

13 Kumaria, A. et al. (2006b). Multimedia big data computing and Internet of Things applications: a taxonomy and process model. *Journal of Network and Computer Applications* 124: 169–195.

14 Kolajo, T. and Daramola, O. (2006b). Big data stream analysis: a systematic literature review. *Journal of Big Data* 6 (21): 169–195.

15 Wang, Z., Mao, S., Yang, L., and Tang, P. (2018). A survey of multimedia big data. 15 (1): 155–176.

16 Dhananjay, S., Kumar, P., and Ashok, A. (2006b). Introduction to multimedia big data computing for IoT. *(eds) Multimedia Big Data Computing for IoT Applications. Intelligent Systems Reference Library* 163: 1–6.

17 Al-Fuqaha, A., Guizani, M., Mohammadi, M. et al. (2015). Internet of Things: a survey on enabling technologies, protocols, and applications. *IEEE Communication Surveys and Tutorials* 17 (4): 2347–2376.

18 Brynjolfsson, E. and Mitchell, T. (2017). What can machine learning do? Workforce implications. *Science* 358 (6370): 1530–1534.

19 Manyika, J., Chui, M., Bughin, J. et al. (2013). Disruptive technologies: advances that will transform life, business, and the global economy. McKinsey Global Institute, M 1, 2013 | Report, 180.

20 Mohammadi, M., Al-Fuqaha, A., Sorour, S., and Guizani, M. (2018). Deep learning for IoT big data and streaming analytics: a survey. *Science* 20 (4): 2923–2960.

21 Tausifa, J.S. and Ahsan, C.M. (2019). Deep learning for Internet of Things data analytics. *Procedia Computer Science* 163 (4): 381–390.

22 Cui, L., Yu, F.R., and Yan, Q. (2016). When big data meets software-defined networking: SDN for big data and big data for SDN. *IEEE Network* 30 (1): 58–65.

23 Lin, B.-S.P., Lin, F.J., and Tung, L.-P. (2016). The roles of 5G mobile broadband in the development of IoT, big data, cloud and SDN. *Communications and Network* 8: 9–21.

24 Wang, G., Ng, T.E., and Shaikh, A. (2012). Programming your network at runtime for big data applications. *Proceedings of the 1st Workshop Hot Topics in Software Defined Networks*, pp. 103–108, August 2012.

25 Monga, I., Pouyoul, E., and Guok, C. (2012). Software-defined networking for big-data science – architectural models from campus to the WAN. *Proceedings of the High Performance Computing, Networking, Storage and Analysis*, pp. 1629–1635, November 2012.

26 Zhang, H., Dong, L., Gao, G. et al. (2020). DeepQoE: a multimodal learning framework for video quality of experience (QoE) prediction. *IEEE Transactions on Multimedia* 22 (12): 3210–3223.

27 Ahmad, A., Mansoor, A.B., Barakabitze, A.A. et al. (2021). Supervised learning based QoE prediction of video streaming in future networks: a tutorial with comparative study. *IEEE Communications Magazine* 59 (11): 88–94.

3

Quality of Experience Management of Multimedia Streaming Services

This chapter addresses the Quality of Experience (QoE) management of multimedia streaming services including its definition, QoE modeling, and assessment: metrics and models, QoE monitoring and measurement, and QoE optimization and control in the context of 5G networks.

3.1 Quality of Experience (QoE): Concepts and Definition

The quality of service (QoS) concept has been used in the past for the Video Quality Assessment (VQA) for the Internet-based video streaming. QoS metric also measured the quality and performance of network services such as video streaming services. However, the QoS concept is limited and centered on the network parameters including packet loss, delay, jitter, etc. QoS does not include the context factors for evaluating the end-to-end aspect of audiovisual streaming systems quality. To overcome the shortcomings of the QoS concept, the user-centric concept [1] of QoE was proposed in 2007 and complemented as a user-centric approach. The QoE-centric concept is comprehensively defined in [1] as follows:

the degree of delight or annoyance of the user of an application or service. It results from the fulfillment of his or her expectations with respect to the utility and/or enjoyment of the application or service in the light of the user's personality and current state, expectation or perception. QoE considers the user's subjectivity toward a specific service. Figure 3.1 shows a blue print of the QoE concept where an end-user is subjected to perceive the video quality from Youtube while watching a movie. In the context of future networks such as 5G and 6G, QoE will be widely used as a service provisioning quality metric that is centered on the management of emerging multimedia services. For example, solutions QoE-driven and energy-aware video adaptation and video quality management [2] in future 5G networks have been proposed [3] (Figure 3.1).

Multimedia Streaming in SDN/NFV and 5G Networks: Machine Learning for Managing Big Data Streaming, First Edition. Alcardo Barakabitze and Andrew Hines.

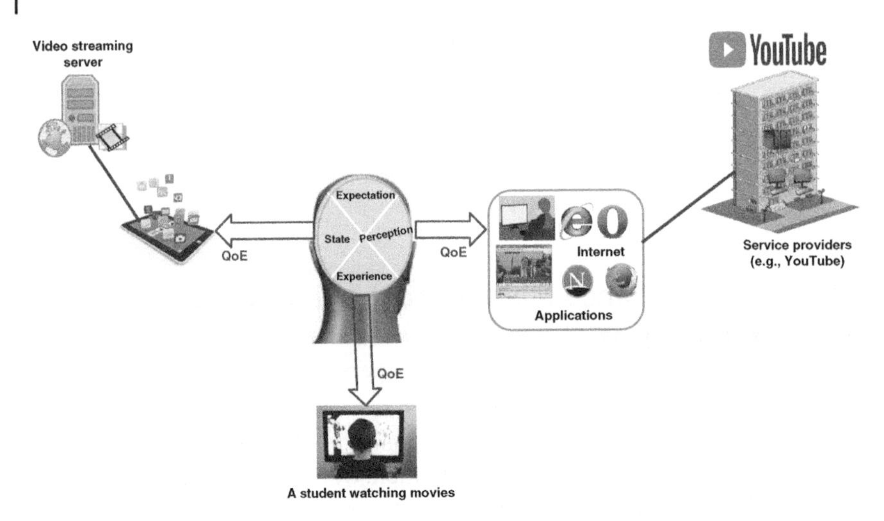

Figure 3.1 Illustration of the QoE concept. Source: Mojzagrebinfo/Pixabay.

3.1.1 Quality of Experience Influence Factors

In end-to-end communication networks (e.g. from service providers to consumers), the QoE is influenced by several factors which can be categorized into three parts: Context Influence Factors (IFs), System Ifs, and Human IFs [3, 4]. The context IFs include the temporal information such as playing time and duration of a video, a person's movements, physical location, and space where the streaming service is being delivered or taking place. The equipment brand, or subscription type (gold, silver, or bronze), and the economic costs for the service can also influence the way customers perceive the received QoE. The system IFs describe the technical characteristics that produce the quality of an application or a service. The system IFs include attributes related to the network, media, content, or equipment, for example factors such as video resolution, encoding, and frame rate are related to media IFs. The video content type and reliability are related to content Ifs, while device and system specifications and device capabilities indicate the device IFs [4]. The mental and emotional constitution of a person, demographic, and socialeconomic background indicate the human IFs. In addition, users' gender, age, mood, attention, and users' personal and cultural traits describe the human IFs [5]. The IFs can be used for video streaming quality assessment. Although these factors can be used to access the video quality, they cannot accurately reflect to the QoE. The next subsection provides a discussion of some QoE assessment methods regarding subjective tests, objective quality models, and data-driven analysis models for quantifying the end-users' video quality.

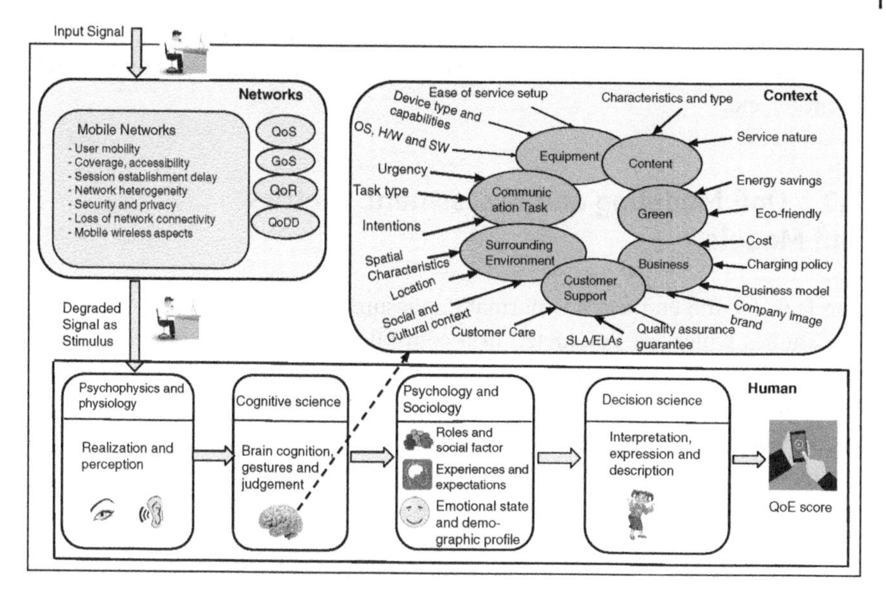

Figure 3.2 QoE influence factors. Source: Danheighton/Adobe Stock.

QoE covers all factors in the end-to-end communication system (e.g. from service providers or operators to clients). These factors can play a great role in influencing the user's experience on the delivered service. As shown in Figure 3.2, these factors can be categorized into three categories, namely System IFs, context IFS, and human IFs [4]. The system IFs describe technical characteristics that produces quality of an application or a service. The system IFs are related to content, media, network, or device. For example in video streaming scenarios, the video content type and content reliability are related to content IFs, while factors such as media configurations, encoding, resolution, and frame rate are related to media IFs. System and equipment specifications and device capabilities represent the device IFs. Most of the experimental results from subjective tests by mobile users watching the same video sequence in the same environment can be highly different. This is generally caused by device-related IFs that includes device capabilities (e.g. display screen, battery), equipment specifications (e.g. mobility), and system specifications (e.g. personalization) at the terminal end [4]. The context IFs include any characteristics information that may influence QoE. For instance, QoE can be influenced by temporal information (e.g. video playing time, video duration), physical location, movements, or space where the service is being provided. The economic factors in terms of cost, equipment brand, or subscription type may affect the received QoE at the end user. The human IFs describe factors that influence how the user perceives or feels the service quality delivered to him or her. It includes emotional and mental constitution of human, demographic and socialeconomic background, the user's personal and cultural traits, user's gender, age, visual/auditory acuity, mood, attention, etc.

3.2 QoE Modeling and Assessment: Metrics and Models

The QoE for the end users is normally measured or assessed based on subjective studies which are conducted in a controlled laboratory. The methods and technique used for subjective studies are elaborated in the ITU-T Rec. BT.500, P.910, and P.913 [3], while the plans and methods for improving subjective tests are provided and guided by the Video Quality Expert Group (VQEG). The video and picture research community have been working to develop different measurement and assessment models in an attempt to predict the quality perceived by the human [6–8]. These models are categorized into three groups: (i) subjective quality models, (ii) Objective Quality Models (OQM), and (iii) data-driven analysis models. For subjective quality models, the participants/subjects assess the video quality using the Mean Opinion Score (MOS) using a scale of 1 meaning (poor) to 5 (excellent) [9]. Although subjective quality assessment provides information on the actual video quality that the end user experiences, the method has drawbacks: (i) it is time-consuming and costly, and (ii) subjective studies are not suitable for real-world applications because few IFs can be evaluated because of testing duration and assessors constraints [10–12].

3.2.1 Subjective Quality Models

For subjective quality models, the end user's QoE is measured based on participant's or subject's opinions of the study conducted in the laboratory. The participants to assess the video quality are normally expressed using a MOS on a scale of 1 meaning (bad) to 5 (excellent). Some of the reference methods for conducting the subjective assessment of video quality are provided by the International Telecommunication Union (ITU) [13–15]. The plans for performing subjective tests in the laboratory are provided and guided by the VQEG. The VQEG provides collaborative efforts to improve subjective video quality test methods as well as developing and validating objective video quality metrics. For the industry, the VQEG seeks to improve the understanding of new video technologies and applications. Although MOS studies have served as the basis for analyzing many aspects of video quality, they present several limitations: (i) subjective tests for assessing the video quality require stringent environments with limited participants/subjects, test videos, and test conditions. (ii) The process cannot be automated. (iii) They are very costly and time-consuming to repeat it frequently. (iv) They are impossible to use in real-time quality assessment.

Contingent upon the measure of source data required, they can be ordered as Full Reference (FR) (full source information required), Reduced Reference (RR) (fractional source information required) furthermore, No-reference (NR) (no source information is required). Because of full admittance to source data, FR measurements are usually more precise than RR and NR measurements. An audit and characterization of existing models and measurements proposed for QoE assessment for HTTP Adaptive Streaming (HAS)-based applications is given by Barman and Martini [8], while a study on QoE measurements and evaluation procedures is given in reference. For a diagram of the QoE estimation approaches, we allude the user to the study and instructional exercise paper by Juluri et al. [16]. Table 3.1 presents the most commonly used quality measurements of picture and video quality evaluation.

3.2.2 Objective Quality Models

Objective quality models for assessing video quality were introduced to address the limitations of subjective test methods [3, 29]. Objective methods tests compare mathematically the source and encoded files and deliver a score for each tested video. Note that, different objective methods investigates how the Human Visual System (HVS) processes and perceives the video signals. The most used method is to provide a quantification of the physical difference between the reference and target (distorted) video. The errors are then weighted according to spatial and video temporal features. There are multiple video metrics, such as peak signal-to-noise ratio (PSNR), Structured Similarity Index Metric (SSIM), video quality metric (VQM), and SSIMPlus [3]. The utility of each metric relates to its ability to predict how human eyes would evaluate the files, or the correlation with subjective results. The Mean Squared Error (MSE) is the least squares (L2) norm of the arithmetic difference between two signals. These two signals are distorted and reference signals. MSE is a signal fidelity measure, which provides a quantitative score of similarity or fidelity of two signals. The PSNR is a prominent metric to measure the quality between distorted and the reference signal. These models are commonly used in different QoE management studies such as [3, 7] and usually serve as the benchmark for assessing the video quality. The PSNR-HVS is an extension of PSNR that incorporates properties of the HVS such as contrast perception. The PSNR-HVS [30] is a modification of PSNR that utilizes the contrast sensitivity function and its model basis operates Discrete Cosine Transform (DCT) on 8×8 pixel blocks, the coefficient of which relies on the HVS. The inputs of the model are the original 8×8 pixel block and the corresponding block of the distorted image. The DCT of the difference between

Table 3.1 Some of the commonly used image and video quality assessment models.

Metric	Year	Model Type	Modality
Peak Signal-to-Noise Ratio (PSNR) [17]		FR	Images – Frames
Structural Similarity Index Metric (SSIM) [18]	2004	FR	Images – Frames
Video Multimethod Assessment Fusion (VMAF) [19]	2016	FR	Images – Frames
Visual Information Fidelity (VIF) [20]	2006	FR	Images – Frames
HDR-VDP-2: A calibrated visual metric for visibility and quality predictions in all luminance conditions [21]	2011	FR	HDR images
Video Quality Metric (VQM) [22]	2004	FR	Video
Reduced Reference Entropic Differencing (RRED) [23]	2013	RR	Video
Spatial Efficient Entropic Differencing for Quality Assessment (SpEED-QA) [24]	2017	RR	Video
Blind Image Quality Index (BIQI) [25]	2010	NR	Images – Frames
Blind/Referenceless Image Spatial QUality Evaluator (BRISQUE) [26]	2012	NR	Images – Frames
Naturalness Image Quality Evaluator (NIQE) [27]	2013	NR	Images – Frames
HDR Image GRADient based Evaluator (HIGRADE) [28]	2017	NR	HDR images

the given pixels are calculated and then are reduced by the value of contrast masking [30]. The method can be calculated for overlapping and nonoverlapping blocks as well.

The Structural SIMilarity (SSIM) index [31] measures the similarity between two images. The SSIM index can be viewed as a quality measure of one of the images being compared provided the other image is regarded as of perfect quality [31]. As an extension to SSIM, SSIMplus is an objective full-reference perceptual video QoE index. It scores quality in a range between 0 and 100 that provides real-time prediction of the perceptual quality of a video based on HVS behaviors, video content characteristics (e.g. video resolution and spatial and temporal complexity), display device properties (e.g. resolution, brightness, and screen size), and viewing conditions (angle and viewing distance). SSIMplus in many ways goes far beyond what SSIM can measure with distinctive features such as high accuracy and high speed, straightforward and easy-to-use, device-adaptive, and cross-resolution QoE assessment. It also provides cross-content QoE measurement and detailed quality map during video assessment. SSIMplus may be employed in many application scenarios including the following: (i) video delivery over multimedia communication networks; (ii) live and file-based video QoE monitoring; (iii) benchmarking video encoders and transcoders; (iv) guiding adaptive bit-rate video coding; and (v) enabling smart quality-driven adaptive bit-rate video streaming. The VQM [32] is standardized under ITU-T J.144 as the NTIA General Model. The VQM metric aligns the frames between the reference and distorted video to compensate for frame losses. VQM consists of objective parameters to measure the perceptual effects of several video impairments such as jerky motion, global noise, blurring, block, error blocks, and color distortion [3].

The Video Intrinsic Integrity and Distortion Evaluation Oracle (VIIDEO) [32] is a blind video quality assessment (VQA) model that does not require the use of any information other than the video being evaluated for quality. The VIIDEO can perform quality prediction of distorted videos without any external knowledge about the training videos containing anticipated distortions or human opinions of video quality or original video source. Although VIIDEO seems to perform better that existing blind objective Image Quality Assessment (IQA) models [33], it may fail to represent some video-specific intrinsic characteristics due to the fact that it can capture only the common baseline characteristics of a specific collection of nondistorted content. The disadvantages of objective methods are [3, 15] (i) they do not correlate well with human perception and require high calculation power, and are time-consuming; (ii) they cannot be used for real-time quality assessment. This is so because objective methods work on both the original video sequence and the transmitted/distorted one; and (iii) it is difficult to build an objective model that consider many quality-affecting parameters, especially network parameters [32].

3.2.3 Data-driven Quality Models

Data-driven analysis has been recognized as an approach that can overcome the limitations of objective and subjective methods for measuring and estimating the end user's QoE. The data-driven analysis incorporates user engagement metrics [34] such as the number of watched videos, the viewing time, return rate, and abandonment rate. The QoE metrics such as rebuffering, video bitrate, stall duration, and startup delay which are measurable and quantifiable are normally used to derive the user engagement metrics which are used during data-driven analysis. The data-driven analysis approach can be used to develop efficient QoE prediction models. This is mainly possible using big data and machine learning paradigms which are able to learn from previous data and forecast the end user's video bitrate and quality requirements. Content providers are interested in data-driven analysis methods since their business objectives and revenue generation can be easily met.

3.3 QoE Measurement, Assessment, and Management

To help network operators and service providers to quickly solve the identified problems in the networks through monitoring, service developers, and QoE, experts conduct an E2E QoE assessment, service quality detection, and delimitation. The QoE monitoring and management of multimedia streaming services require information with respect to the main driver of QoE degradation or unsatisfactory QoE levels in the network or end user's side. To achieve this, QoE-related data/information related to specific application/service and its quantification, device capabilities (e.g. screen size, resolution), and QoE-based information inside the network must be monitored, gathered, and measured [3, 35]. Crowdsourcing [36] is another approach that can be used to monitor QoE provisioning in 5G systems. This is done by collecting quality information of communication services on the 5G network and users' mobile terminals. Crowdsourcing can enable systematic verification of users' inputs and derives interval-scale scores that enable subsequent quantitative analysis and QoE provisioning to the end users. With the advancement of network virtualization and data management capabilities, virtualized passive probes (vProbes) [35] installed in the dedicated hardware can monitor the network traffic and test streaming media applications in softwarized networks. The QoE estimation and prediction models are used to estimate or predict the end user's QoE based on the inputs data received from the collected network-level parameters (such as packet loss, delay) and user-level service/application specific quality parameters (video resolution, frame rate, service usability, etc.) [37].

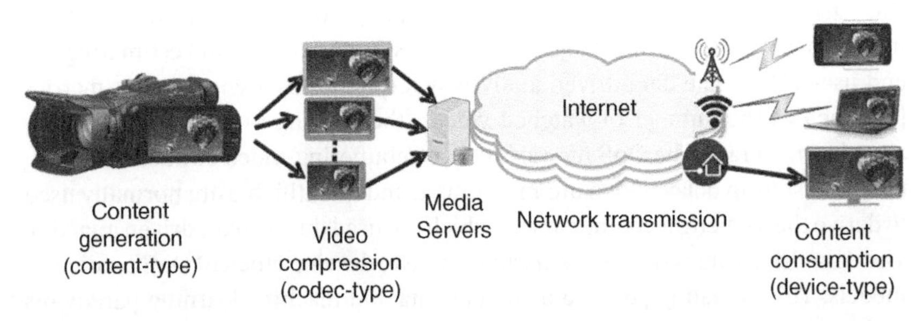

Figure 3.3 Multimedia streaming chain. Source: Barakabitze et al. [3] / with permission of IEEE.

3.4 QoE Optimization and Control of Multimedia Streaming Services

QoE is a measurement of customer satisfaction for experience with services (e.g. browsing, calls, video streaming). The QoE management of multimedia services, as shown in Figure 3.3 entails continuous optimization and dynamic control of relevant mechanisms, from content generation to content consumption, along the service delivery chain. One of the ultimate objectives of QoE management is the maximization of the end users' QoE level through the efficient allocation of available network resources. However, by considering the Figure 3.3, QoE optimization and control is facing different challenging task due to many issues, such as the heterogeneity of multimedia-capable users' devices. Authors in [38] state that the main challenges that arise with regards to QoE optimization and control can answer the following questions: (i) what key quality parameters to optimize and control? (ii) Where to control (e.g. at the client, server and the network side)? (iii) When to perform QoE optimization and control (e.g. during the service, that is, on-line control, or in an off-line fashion [17])? (iv) How often to control and optimize QoE? In order to answer those questions, different studies have been conducted [3] (Figure 3.5).

Based on the 3GPP IP multimedia subsystem architecture (IMS), a QoE-driven services adaption and network optimization method employing resource allocation was proposed in [40]. The proposed QoS/QoE matching and optimization function [41] was used to optimize the overall utility of active sessions and determine the ideal service configurations given the available network resources and operator's QoE service policy. The agreed service profile, which contains multiple service-level parameters, is used to calculate the QoE-optimization process (e.g. frame rate and codec type). The optimization function gathers the required input data before invoking the optimization engine, which executes the QoE optimization methods. Following which, the final QoE is calculated based on the negotiated and agreed-upon service profiles between end users and service providers [41]. Additional 3GPP QoE control mechanisms have been proposed and extensively investigated in [42], in which a QoE estimation function collects a set of QoE Influence Factors (IFs) and operates at the application server. QoE-driven optimization, from the standpoint of network operators, entails network resource management strategies. To provide quality assurance and service control in the network, such approaches are often based on quality-related information received via the QoE monitoring and measuring process.

Mobile networks or MESH networks have employed QoE-driven optimization for resource management. For adaptive multimedia services in LTE networks, some of the strategies employed include cross-layer QoE-driven admission control and resource allocation [43]. Figure 3.4 demonstrates a QoE-driven

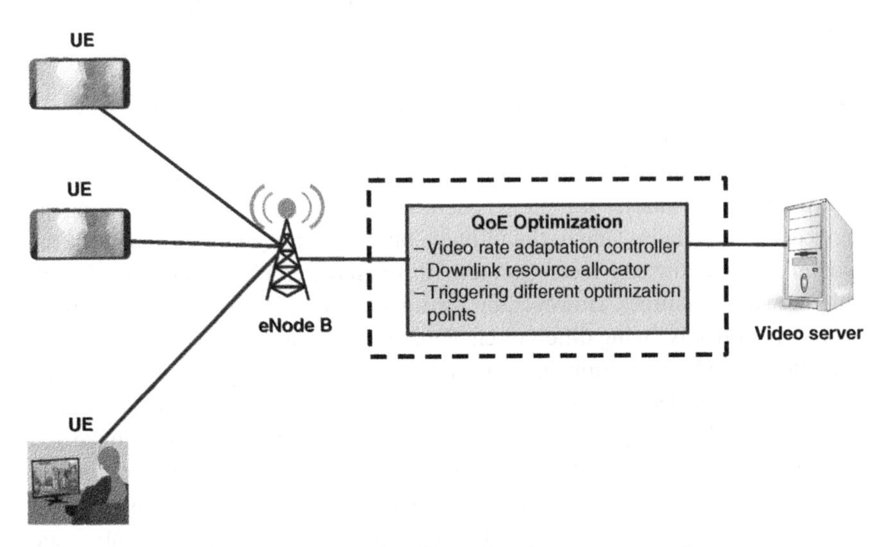

Figure 3.4 QoE optimization of multimedia services in access networks.

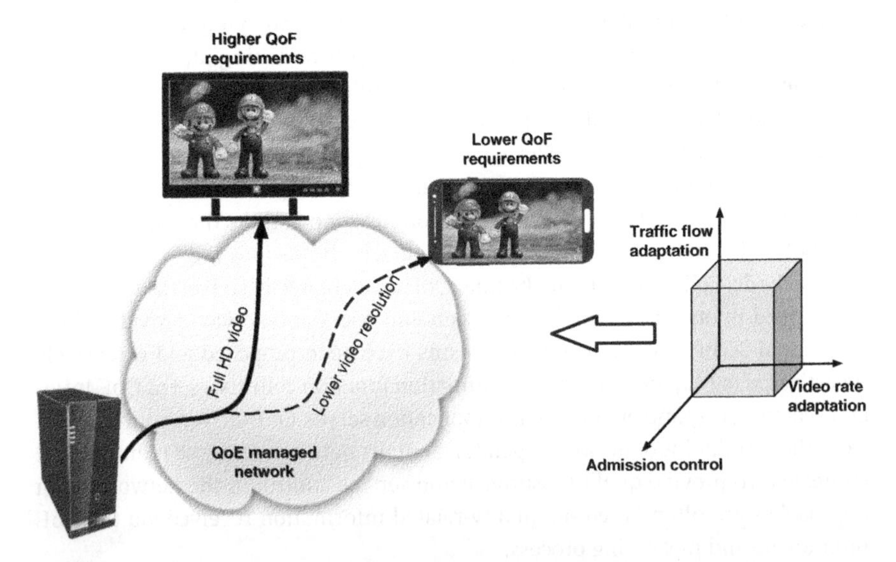

Figure 3.5 QoE in multimedia network. QoE is automatically managed using three different axes, namely, (a) traffic-flow adaptation, (b) admission control, and (c) video rate adaptation. Source: Adapted from [39]; Alexas_Fotos / Pixabay.

resource allocation strategy enabling numerous users to access diverse contents in a wireless environment [44]. The end users' perceived quality is increased by rate adaptation and network resource allocation methods as part of a utility maximization strategy. For video rate adaptation, the QoE optimizer in the core network serves as a controller and a downlink resource allocator. Aquarema is a network resource management mechanism proposed by Staehle et al. [45], which enhances end user QoE in all types of applications for any type of network. QoE degradation is avoided with Aquarema, thanks to network management tools and the interaction of application monitoring tools on the client side. Wamser et al. [46] present the YouTube Performance Monitoring Application (YoMoApp), an Android application that can passively monitor key performance indicators (KPIs) of YouTube adaptive video streaming on end-user smartphones. The client-side monitoring tool provides a quantification of status for running apps and allows for the prediction of end-user QoE. During QoE degradation, a network advisor is also utilized to trigger other resource management solutions. The YoMo App [46] tool, which continuously analyzes the amount of playtime delayed by the YouTube player, allows for such interactivity.

Authors in [47] present a joint optimization strategy for network resource allocation and video quality adaption that fairly maximizes the QoE of video clients. Another key consideration when maximizing the end user's QoE has been minimizing energy consumption, particularly in the context of mobile services [48]. Tao et al. [49] present an energy-efficient video QoE optimization approach with a focus on DASH over wireless/mobile networks. In order to maximize QoE, the proposed approach efficiently allocates network resources and makes optimal bitrate selection for clients. Bouten et al. [50] and [51] present in-network quality optimization agents that use sampling-based measuring techniques to monitor the available throughput. As a result, each client's quality is optimized based on a HAS QoE statistic. Xu et al. [52] explore the buffer starvation of video streaming services, whereas Triki et al. [53] suggest a dynamic closed-loop QoE optimization for video adaption and delivery. To improve end user QoE of video streaming services, the authors take advantage of the trade-off between startup/rebuffering delay and starvation. In small-cell networks, Zhao and coworkers [54] present a cooperative transmission technique for video transmission that decreases video freezes and increases QoE. When there is a large number of active users on the network, the system performance improves dramatically. In this case, the greedy method uses distributed caching to transmit video-file portions.

In addition, a central QoE management entity that monitors and regulates the end-QoE user's level is introduced in [55]. The proposed architecture shown in Figure 3.6 can collect QoE-related inputs and make network management decisions based on QoE. It is made up of three parts: a QoE monitor, a QoE manager, and a QoE controller. The QoE controller is an interface that allows communication between the central QoE management entity and the underlying network to be synchronized. It gathers the necessary information and sends it to the QoE monitor and management as inputs. The QoE monitor is used to estimate and report the QoE per-flow to the QoE management. It classifies traffic using statistical analysis and various built-in "QoE models" based on the type of traffic/service. The QoE manager is in charge of overseeing the customer's viewing experience. It estimates the end users' QoE based on inputs from the QoE controller and determine the network's state during a video streaming session. To improve end user QoE while maintaining network stability and availability, and eliminate service interruptions during video streaming, two or more QoE controllers might be employed.

To achieve QoE control and administration of multimedia services, Latré et al. [39, 56] use admission control, traffic-flow adaptation or video rate adaptation mechanisms, as shown in Figure 3.5. The traffic-flow adaptation modifies the network delivery of a traffic flow by incorporating redundant data into the application data. In order to avoid network congestion, the admission control allows or disallows new connections. As part of the multimedia streaming services provided

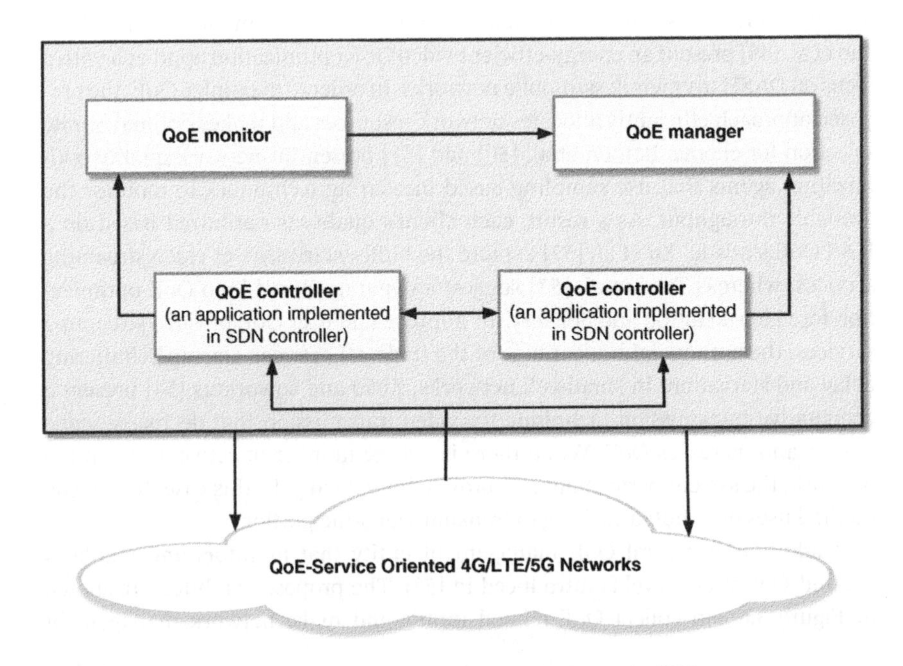

Figure 3.6 QoE-based management in mobile cellular networks [55].

from the server to the clients, the video rate adaptation technique modifies the video quality level. Although the authors in [39] show that these strategies can increase end-user QoE, the exchange of information between layers adds to the deployment complexity.

3.4.1 Customer Experience Management and QoE Monitoring

Customer Experience Management (CEM) is an approach designed to focus on methodologies and procedures that control and manage the end user's QoE [57]. Specifically, the CEM provides a monitoring architecture that can manage and optimize the end user's QoE. The QoE monitoring is set to monitor the input QoE parameters using specific configurations as defined in the ELAs (e.g. active or passive monitoring strategy). It is responsible for gathering information from different parts of the network (e.g. RANs, user's terminal) which can be relevant for the delivered service. The E2E QoE monitoring on 5G systems needs powerful data aggregation nodes/engines and efficient tools for E2E mapping of services in order to enhance and guarantee the QoE level of offered services.

CEM is an approach that focuses on methodologies and procedures to control and manage the end user's QoE [57]. An efficient QoE monitoring and management system that retrieve information related to each QoE IFs over 5G softwarized networks is of great importance. The QoE monitoring architecture shown in Figure 3.7 can monitor in real time the E2E service KPIs and estimate the customer experience by using 5G network performance and monitoring tools. This can be designed to have different monitoring points and a realization between service providers and network operators to (i) provide in real time, the QoE monitoring support for large variation traffic for 5G users in business, residential, and public areas, (ii) have visibility and full access of E2E connectivity in all network segments such as transit ISPs, content server side networks, 5G core network, aggregation networks, and customer's premises network, and (iii) controlling the QoE traffic flows dynamically based on changing network conditions.

The challenge of performing root cause analysis (e.g. from a network provider perspective) in determining the cause of measured QoE degradation is the lack of E2E visibility which is one of the requirements for QoE monitoring. Network providers are still more comfortable to monitor QoS than QoE, while customer satisfaction is largely driven by QoE and not QoS. Given the dynamic nature of softwarized and virtualized networks, and the inability of current monitoring tools to assure virtualized-based services, network providers have less ability to monitor QoE or the service experience. That way, network providers have to perform per-application level quality assurance for different services and meet user's QoE requirements and the need for automated processes in softwarized and virtualized 5G and beyond networks.

The QoE monitoring and management solution shown in Figure 3.7 consists of major functions responsible to collect data/information related to system IFs, human Ifs, and context IFs. It takes into account the characteristics of a user (e.g. preferences), the characteristics that determine the content (e.g. content type), the network (e.g. data transmission over the network) and device (e.g. system specification), and the factors that describe the user's environment (e.g. location). Information related to system IFs such as content types can be retrieved by using active and passive probes monitoring approaches. Active monitoring can be used to initiate a service connection or send and receive data to measure 5G network transmission quality within a specified time. For example, an active probe can be implemented to set up a 5G test Voice over Internet Protocol (VoIP) phone call to another active probe.

The first monitoring point (MP1) shown in Figure 3.7 indicates the data acquisition point and ensures that video streams for the end users are generated without errors. Service providers at this measurement point also ensure that the agreed QoS/QoE policy (video bitrate, encoding rate, QoS parameters, etc.,) is met based on the subscription and QoE preferences of the end users. The MP2 shows the data measurement point where service providers check for QoS/QoE policies and collect information on quality and performance parameters of video streaming services. MP3 indicates the measurement and monitoring point where IP network-related parameters (packet delay variation, packet loss ratio, etc.,) are measured. MP4 shows the boundary between the access network and end users' premises where information regarding packet loss ratio, IP network parameters, and reliability delivery of data can be measured and collected. The MP5 defines the QoE point at the end-user devices where the perceived video quality can be evaluated using application-specific information or the device characteristics.

QoE monitoring can be done using crowdsourcing or virtualized probes (in SDN/NFV-based environments) by providing insights and real-time data on the state of the network and the end-user's QoE. Crowdsourcing [58] is another approach that can be used to monitor QoE provisioning in 5G systems. This is done by collecting quality information of communication services on the 5G network and users' mobile terminals. Crowdsourcing can enable systematic verification of user's inputs and derives interval-scale scores that enable subsequent quantitative analysis and QoE provisioning to the end users. In a virtualized network environment, virtualized passive probes (vProbes) [35] installed in the dedicated hardware can be used for monitoring traffic and testing of applications in 5G softwarized networks. Due to the increase in encrypted video traffic, QoE-related metrics (Start-up delay and rebuffering (stalling) from encrypted traffic of video streaming is a topic of interest from the academia and industry [59, 60]. Since 2016, a majority of YouTube traffic has been encrypted with a combination of quick UDP internet connections (QUIC) and HTTPS. Since then,

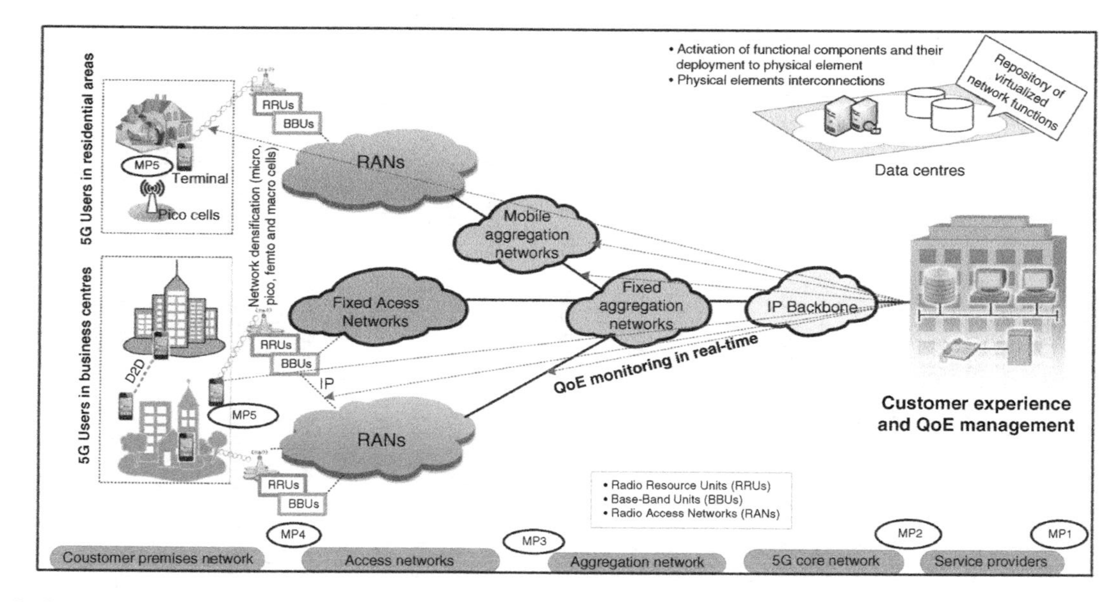

Figure 3.7 QoE monitoring for large traffic variations in business, residential, and public areas in 5G networks.

content providers have been increasingly adopting E2E encryption. Orsolic et al. [61] and Dimopoulos et al. [62] use machine learning to classify Youtube QoE based on encrypted network traffic. Authors predict QoE metrics of stalls, and its variations using network traffic measurements (e.g. interarrival time, RTT, throughput, and bandwidth delay product).

Wehner et al. [60] propose a web QoE-monitoring approach for encrypted network traffic through time series modeling, while Khairi et al. [59] introduce a novel architecture for 5G networks that can enhance the end user's QoE monitoring and management by integrating SDN and enhanced telecom operations map (eTOM). The proposed architecture can enable dynamic network monitoring for 5G networks by exploiting SDN functionalities and satisfy the ELA/SLA constraints by enforcing suitable strategies to adjust network parameters. Khokhar et al. [63] apply ML under realistic network conditions to build ML-based models (classification and regression) that estimate QoE (the stalling events, the startup delay, the quality switches, the spatial resolution of playout, and the MO) from encrypted video traces using network level measurements. Huet et al. [64] use ML to build data-driven models that estimate well-known application-level Web browsing QoS metrics (e.g. Page Load Time and SpeedIndex) from encrypted streams of network traffic. Authors measure both the application-level parameters (AppQoS metrics) and the network-level parameters (packet-level traces) after collecting a large dataset of Web page load in a controlled experiment.

Authors in [65] develop Request, a real-time QoE metric (buffer warning, video state, and video resolution) that identifies video and audio segments from the IP headers of encrypted network traffic. A large YouTube dataset which consists of different video assets are collected and delivered through various Wi-Fi changing network conditions. Request achieves prediction accuracy in terms of video state, buffer low warning, and video resolution compared to the baseline approach [65]. Although the E2E encryption of the video streaming services provides challenges to ISPs to monitor and estimate video quality, efforts have been made to develop solutions regarding QoE monitoring and management in SDN platforms [59], methodology for detecting video streaming QoE issues from encrypted traffic [62], QoE estimation models, and QoE inferring video streaming quality from encrypted traffic [66]. This development provides a significant capability for ISPs and MNOs to monitor and optimize the encrypted video traffic on 5G and beyond networks. It is worth mentioning that using the proposed approaches, [62–66] MNOs and ISPs can optimize, balance, and prioritize encrypted video/network traffic and maintain a satisfactory QoE for video streaming services.

3.5 Conclusion

For QoE-based user-centric service management of streaming services, QoE-based modeling, monitoring, control, and management have become hot topics in the multimedia communication research community. This chapter presents the QoE-aware management, optimization, and control of multimedia streaming services consists of three major components: (a) QoE modeling and assessment; (b) QoE monitoring and measurement; and (c) QoE control and optimization. The QoE modeling and assessment provides mechanisms for developing QoE models related to assess the KQIs of multimedia streaming services. The QoE monitoring and measurement stage provides mechanisms that can monitor the KQIs. It also measures the QoE of users based on the QoE predictive models. The QoE optimization and control performs primarily resource optimization at the client, network, or server side by controlling important actions/states based on the QoE measurements.

Bibliography

1 Callet, P.L., Möller, S., and Perkis, A. (2012). Qualinet white paper on definitions of quality of experience (2012). *European Network on Quality of Experience in Multimedia Systems and Services (COST Action IC 1003)*, 1.2, March 2012.

2 Awobuluyi, O., Nightingale, J., Wang, Q., and Alcaraz-Calero, J.M. (2015). The QoE implications of ultra-high definition video adaptation strategies. *2015 IEEE International Conference on Computer and Information Technology; Ubiquitous Computing and Communications; Dependable, Autonomic and Secure Computing; Pervasive Intelligence and Computing.*

3 Barakabitze, A.A., Barman, N., Ahmad, A. et al. (2020). QoE management of multimedia services in future networks: a tutorial and survey. *IEEE Communication Surveys and Tutorials* 22 (1): 526–565.

4 Brunnström, K., Beker, S.A., De Moor, K. et al. (2013). Qualinet white paper on definitions of quality of experience.

5 Reiter, U., Brunnström, K., De Moor, K. et al. (2014). *Factors Influencing Quality of Experience: In Quality of Experience: Advanced Concepts, Applications and Methods*, 55–72. Springer International Publishing.

6 Barakabitze, A.A., Mkwawa, I.-H., Hines, A. et al. (2020). QoEMultiSDN: management of multimedia services using MPTCP/SR in softwarized and virtualized networks. *IEEE ACCESS* 8: 1–1.

7 Skorin-Kapov, L., Varela, M., Hoßfeld, T., and Chen, K.-T. (2018). A survey of emerging concepts and challenges for QoE management of multimedia

services. *ACM Transactions on Multimedia Computing, Communications, and Applications (TOMM)* 14 (2s): 2–28.

8 Barman, N. and Martini, M.G. (2019). QoE modeling for HTTP adaptive video streaming-a survey and open challenges. *IEEE Access* 7:30831–30859. https://doi.org/10.1109/ACCESS.2019.2901778.

9 Fan, Z. and Jiang, T. (2014). Subjective quality assessment of mobile 3D videos. *6th International Workshop on Quality of Multimedia Experience (QoMEX)*, pp. 226–231, Singapore, September 2014.

10 Staelens, N., De Meulenaere, J., Claeys, M. et al. (2014). Subjective quality assessment of longer duration video sequences delivered over HTTP adaptive streaming to tablet devices. *IEEE Transactions on Broadcasting* 60 (4): 707–714.

11 Baraković, S. and Skorin-Kapov, L. (2013). Survey and challenges of QoE management issues in wireless networks. *Journal of Computer Networks and Communications* 2013 165146:1–165146:28.

12 Kua, J., Armitage, G., and Branch, P. (2017). A survey of rate adaptation techniques for dynamic adaptive streaming over HTTP. *IEEE Communication Surveys and Tutorials* 19 (3): 1842–1866.

13 Chen, C., Inguva, S., Rankin, A., and Kokaram, A. (2016). A subjective study for the design of multi-resolution ABR video streams with the VP9 codec. *SPIE Electronic Imaging, Human Visual Perception*, San Francisco, California, USA, January 2016.

14 Möller, S., Antons, J.N., Beyer, J. et al. (2015). Towards a new ITU-T recommendation for subjective methods evaluating gaming QoE. *2015 Seventh International Workshop on Quality of Multimedia Experience (QoMEX)*, pp. 1–6, Pylos-Nestoras, Greece, May 2015.

15 Seshadrinathan, K., Soundararajan, R., Bovik, A.C.a.C. et al. (2010). Study of subjective and objective quality assessment of video. *IEEE Transactions on Image Processing* 19 (6): 1427–1441.

16 Juluri, P., Tamarapalli, V., and Medhi, D. (2016). Measurement of quality of experience of video-on-demand services: a survey. *IEEE Communication Surveys and Tutorials* 18 (1): 401–418. https://doi.org/10.1109/COMST.2015.2401424.

17 Girod, B. (1993). *What's Wrong with Mean-squared Error?* Digital Images and Human Vision. ACM, pp. 207–220. http://dl.acm.org/citation.cfm?id=197765.197784 (accessed 11 August 2022).

18 Wang, Z., Bovik, A.C., Sheikh, H.R., and Simoncelli, E.P. (2004). Image quality assessment: from error visibility to structural similarity. *IEEE Transactions on Image Processing* 13 (4): 600–612.

19 Netflix (2016). Toward a practical perceptual video quality metric. https://medium.com/netflix-techblog/toward-a-practical-perceptual-video-quality-metric-653f208b9652 (accessed 06 March 2018).

20 Sheikh, H.R. and Bovik, A.C. (2006). Image information and visual quality. *IEEE Transactions on Image Processing* 15 (2): 430–444.

21 Mantiuk, R., Kim, K.J., Rempel, A.G., and Heidrich, W. (2011). HDR-VDP-2: a calibrated visual metric for visibility and quality predictions in all luminance conditions. *ACM Transactions on Graphics* 30 (4): 40:1–40:14.

22 Pinson, M.H. and Wolf, S. (2004). A new standardized method for objectively measuring video quality. *IEEE Transactions on Broadcasting* 50 (3): 312–322. https://doi.org/10.1109/TBC.2004.834028.

23 Soundararajan, R. and Bovik, A.C. (2013). Video quality assessment by reduced reference spatio-temporal entropic differencing. *IEEE Transactions on Circuits and Systems for Video Technology* 23 (4): 684–694. https://doi.org/10.1109/TCSVT.2012.2214933.

24 Bampis, C.G., Gupta, P., Soundararajan, R., and Bovik, A.C. (2017). SpEED-QA: spatial efficient entropic differencing for image and video quality. *IEEE Signal Processing Letters* 24 (9): 1333–1337.

25 Moorthy, A.K. and Bovik, A.C. (2010). A two-step framework for constructing blind image quality indices. *IEEE Signal Processing Letters* 17 (5): 513–516. https://doi.org/10.1109/LSP.2010.2043888.

26 Mittal, A., Moorthy, A.K., and Bovik, A.C. (2012). No-reference image quality assessment in the spatial domain. *IEEE Transactions on Image Processing* 21 (12): 4695–4708. https://doi.org/10.1109/TIP.2012.2214050.

27 Mittal, A., Soundararajan, R., and Bovik, A.C. (2013). Making a "completely blind" image quality analyzer. *IEEE Signal Processing Letters* 20 (3): 209–212. https://doi.org/10.1109/LSP.2012.2227726.

28 Kundu, D., Ghadiyaram, D., Bovik, A.C., and Evans, B.L. (2017). No-reference quality assessment of tone-mapped HDR pictures. *IEEE Transactions on Image Processing* 26 (6): 2957–2971. https://doi.org/10.1109/TIP.2017.2685941.

29 Chen, C., Choi, L.K., De Veciana, G. et al. (2014). Modeling the time-varying subjective quality of HTTP video streams with rate adaptations. *IEEE Transactions on Image Processing* 23 (5): 2206–2221.

30 Gupta, P., Srivastava, P., Bhardwaj, S., and Bhateja, V. (2011). A modified PSNR metric based on HVS for quality assessment of color images. *International Conference on Communication and Industrial Application*, Uttar Pradesh, Lucknow, December 2011.

31 Wang, Z., Bovik, A.C., Sheikh, H.R., and Simoncelli, E.P. (2004). Image quality assessment: from error visibility to structural similarity. *IEEE Transactions on Image Processing* 4 600–612.

32 Pinson, M.H. and Wolf, S. (2004). A new standardized method for objectively measuring video quality. *IEEE Transactions on Broadcasting* 50 (3): 312–322.

33 Ding, K., Ma, K., Wang, S., and Simoncelli, E.P. (2020). Comparison of image quality models for optimization of image processing systems. *International Journal of Computer Vision*, May 2020.

34 Wassermann, S., Wehner, N., and Casas, P. (2018). Machine learning models for youtube QoE and user engagement prediction in smartphones. *ACM SIGMETRICS Performance Evaluation Review* 46 (3): 155–158.

35 Tselios, C. and Tsolis, G. (2016). On QoE-awareness through virtualized probes in 5G networks. *2016 IEEE 21st International Workshop on Computer Aided Modelling and Design of Communication Links and Networks (CAMAD)*, pp. 159–164. October 2016.

36 Hoßfeld, T., Seufert, M., Sieber, C., and Zinner, T. (2014). Assessing effect sizes of influence factors towards a QoE model for HTTP adaptive streaming. *2014 6th International Workshop on Quality of Multimedia Experience (QoMEX)*, pp. 111–116, Singapore, September 2014.

37 Robitza, W., Ahmad, A., Kara, P.A., Atzori, L. et al. (2017). Challenges of future multimedia QoE monitoring for internet service providers. *Multimedia Tools and Applications* 3 22243–22266.

38 Barakovi, S. and Skorin-Kapov, L. (2013). Survey and challenges of QoE management issues in wireless networks. *Journal of Computer Networks and Communication* 2013 1–28.

39 Latré, S., Simoens, P., and De Turck, F. (2012). Autonomic quality of experience management of multimedia networks. *IEEE Network Operations and Management Symposium*, pp. 872–879, April 2012.

40 Skorin-Kapov, L. and Matijasevic, M. (2009). Modeling of a QoS matching and optimization function for multimedia services in the NGN. *Proceedings of the 12th IFIP/IEEE International Conference on Management of Multimedia and Mobile Networks and Services: Wired-Wireless Multimedia Networks and Services Management (MMNS'09)*, pp. 55–64, Italy, October 2009.

41 Ivec, K., Matijasevic, M., and Skorin-Kapov, L. (2011). Simulation based evaluation of dynamic resource allocation for adaptive multimedia services. *Proceedings of the 7th International Conference on Network and Service Management (CNSM'11)*, pp. 1–8, Paris, France, October 2011.

42 Sterle, J., Volk, M., Sedlar, U. et al. (2011). Application based NGN QoE controller. *IEEE Communications Magazine* 1 (49): 92–101.

43 Ivesic, K., Skorin-Kapov, L., and Matijasevic, M. (2014). Cross-layer QoE-driven admission control and resource allocation for adaptive multimedia services in LTE. *Journal of Network and Computer Applications* 46 (C): 336–351.

44 Thakolsri, S., Kellerer, W., and Steinbach, E. (2011). QoE-based cross-layer optimization of wireless video with unperceivable temporal video quality fluctuation. *2011 IEEE International Conference on Communications (ICC)*, pp. 1–6, Kyoto, Japan, June 2011. https://doi.org/10.1109/icc.2011.5963296.

45 Staehle, B., Hirth, M., Pries, R. et al.. Aquarema in action: improving the YouTube QoE in wireless mesh networks. *Baltic Congress on Future Internet and Communications*, pp. 33–40, Riga, Latvia, February 2011. https://doi.org/10.1109/BCFIC-RIGA.2011.5733220.

46 Wamser, F., Seufert, M., Casas, P. et al. (2015). YoMoApp: a tool for analyzing QoE of youtube http adaptive streaming in mobile networks. *2015 European Conference on Networks and Communications (EuCNC)*, pp. 239–243. IEEE.

47 Joseph, V. and Veciana, G. (2014). Nova: QoE-driven optimization of dash-based video delivery in networks. , *2014 Proceedings IEEE INFOCOM*, pp. 82–90. IEEE.

48 Hoque, M.A., Siekkinen, M., and Nurminen, J.K. (2014). Energy efficient multimedia streaming to mobile devices: a survey. *IEEE Communication Surveys and Tutorials* 16 (1): 579–597.

49 Tao, L., Gong, Y., Jin, S., and Quan, Z. (2016). Energy efficient video QoE optimization for dynamic adaptive HTTP streaming over wireless networks. *IEEE International Conference on Communication Systems*, pp. 1–6, May 2016.

50 Bouten, N., Schmidt, R.O., Famaey, J. et al. (2015). QoE-driven in-network optimization for Adaptive Video Streaming based on packet sampling measurements. *Computer Networks* 81: 96–115.

51 Bouten, N., Famaey, J., and Latre, S. (2012). QoE optimization through in-network quality adaptation for HTTP Adaptive Streaming. *8th International Conference on Network and Service Management (CNSM) and 2012 Workshop on Systems Virtualization Management (SVM)*, Las Vegas, NV, USA, June 2012.

52 Xu, Y., Altman, E., El-Azouzi, R. et al. (2014). Analysis of buffer starvation with application to objective QoE optimization of streaming services. *IEEE Transactions on Multimedia* 16 (3): 813–827. https://doi.org/10.1109/TMM.2014.2300041.

53 Triki, I., El-Azouzi, R., Haddad, M. et al. (2017). Learning from experience: a dynamic closed-loop QoE optimization for video adaptation and delivery. *2017 IEEE 28th Annual International Symposium on Personal, Indoor, and Mobile Radio Communications (PIMRC)*, pp. 1–5, October 2017. https://doi.org/10.1109/PIMRC.2017.8292500.

54 Liu, X., Zhao, N., Yu, F.R. et al. (2018). Cooperative video transmission strategies via caching in small-cell networks. *IEEE Transactions on Vehicular Technology* 67 (12): 12204–12217.

55 Liotou, E., Tsolkas, D., Passas, N., and Merakos, L. (2015). Quality of experience management in mobile cellular networks: key issues and design challenges. *IEEE Communications Magazine* 53 (7): 145–153.

56 Latré, S., Simoens, P., De Vleeschauwer, B. et al. (2009). An autonomic architecture for optimizing QoE in multimedia access networks. *Computer Networks* 53 (10): 1587–1602.

57 Morais, A., Cavalli, A.R., Tran, H.A. et al. (2012). Managing customer experience through service quality monitoring. *Future Network & Mobile Summit 2012 Conference Proceedings*, pp. 1–9. July 2012.

58 Hossfeld, T., Hirth, M., Redi, J. et al. (2014). Best practices and recommendations for crowdsourced QoE - Lessons learned from the Qualinet Task Force Crowdsourcing. *European Network on Quality of Experience in Multimedia Systems and Services (COST Action IC 1003 Qualinet*, pp. 1–27. October 2014.

59 Khairi, S., Raouyane, B., and Bellafkih, M. (2020). Novel QoE monitoring and management architecture with eTOM for SDN-based 5G networks. *Cluster Computing* 23: 1–12.

60 Wehner, N., Seufert, M., Schuler, J. et al. (2020). Improving web QoE monitoring for encrypted network traffic through time series modeling. *Conference: 2nd Workshop on AI in Networks and Distributed Systems*, November 2020.

61 Orsolic, I., Pevec, D., Suznjevic, M., Skorin-Kapov, L. et al. (2017). A machine learning approach to classifying youtube QoE based on encrypted network traffic. *Multimedia Tools and Applications* 76: 22267–22301.

62 Dimopoulos, G. et al. (2016). Measuring video QoE from encrypted traffic. *Proceedings of the 2016 Internet Measurement Conference*, pp. 513–526. November 2016.

63 Khokhar, M. et al. (2019). From network traffic measurements to QoE for internet video. *IFIP Networking Conference*, May 2019.

64 Huet, A. et al. (2020). Revealing QoE of web users from encrypted network traffic. *IFIP Networking Conference (Networking)*, July 2020.

65 Gutterman, C. et al. (2020). Requet: Real-time QoE metric detection for encrypted youtube traffic. *ACM Transactions on Multimedia Computing, Communications, and Applications* 16 (2s): 1–28.

66 Bronzino, F., Schmitt, P., Ayoubi, S. et al. (2019). Inferring streaming video quality from encrypted traffic: practical models and deployment experience. *Proceedings of the ACM on Measurement and Analysis of Computing Systems* 3 (2019): 1–25.

4

Multimedia Streaming Services Over the Internet

With the rise in streaming video traffic, it is become more important than ever to take advantage of numerous elements throughout the multimedia supply chain in order to maximize video service delivery while keeping end users' QoE in mind. This chapter presents a comprehensive discussion on HTTP Adaptive Streaming (HAS) solutions over the Internet. In particular, it presents the history of Internet-based video streaming followed by the Dynamic Adaptive Streaming over HTTP (DASH). It also provides the Server and Network Assisted DASH (SAND) architecture and its implementation in the context of future softwarized networks. Furthermore, this chapter presents multimedia delivery chain and service management issues focusing on internet exchange points (ISPs), over-the top provider (OTTP), transit providers, contend delivery networks (CDNs), and IXPs.

4.1 Internet-Based Video Streaming: An Overview

4.1.1 A Brief History of Internet-Based Video Streaming

For many decades, the biggest growth area in Internet usage has been video streaming. Today, users are enjoying high video resolution and 3D videos on-demand using different devices such as tablets, laptops, smartphones, television, and video players. The media have to transfer the video from the streaming server using two different protocols, namely (i) connection-oriented protocols (e.g. real-time messaging protocol (RTMP/TCP)), and (ii) connectionless protocols (e.g. real-time transport protocol (RTP/UDP)). The real-time streaming protocol (RTSP) is a commonly used protocol for controlling the transmission of video from media servers to clients in multimedia streaming networks. During video streaming, a RTSP first perform streaming session setup and keep the information throughout this streaming session. Protocols such as RTP are responsible for the video delivery from the server to the client. On the video server side, video

Multimedia Streaming in SDN/NFV and 5G Networks: Machine Learning for Managing Big Data Streaming, First Edition. Alcardo Barakabitze and Andrew Hines.

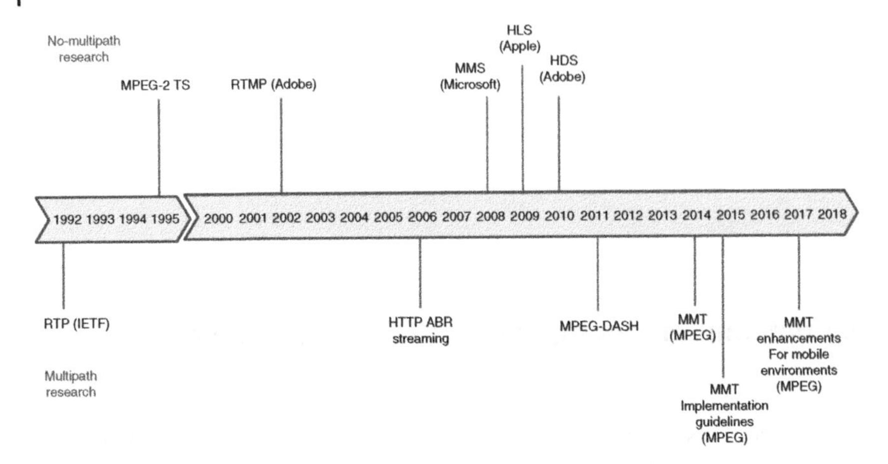

Figure 4.1 History of video streaming.

rate adaptation, and data delivery are performed using the RTP control protocol (RTCP). It is worth mentioning that the RTP is session-oriented such that it provides data for the application to perform: (i) source identification, (ii) packet loss detection and packet re-sequencing, (iii) intramedia synchronization for video playout with jitter buffer, and (iv) intermedia synchronization: between audio and video during a streaming session.

In 1990s, the Internet video streaming witnessed the audio and video improvements in terms of video codecs. Since then, one video server was responsible for the delivery of the video requests to clients through a unicast connections [1]. The scalability of unicast approach for video streaming triggered proposals for new IP multicast protocols which became more scalable when streaming a video to a large number of users. IP multicast system requires that both the network and client devices support the Internet Group Management Protocol (IGMP). IP multicast is commonly used for Internet Protocol Television (IPTV) applications and for the delivery of content in both corporate and private networks [1]. However, the main drawbacks of video multicast are the heterogeneity of devices which have different capabilities especially for delivering QoE to users. This is so because using multicast for video streaming becomes very difficult to find a video bitrate that is appropriate for different hardware capabilities and network resources of multiple users at the same time. Progressive download is another technique used to stream media from a web server to a client such as a video player on a laptop or a mobile phone. The video is actually downloaded to the viewer's computer and stored in a temporary directory. Content starts to play as soon as it is available locally. The progressive video download can support CDN because of massive popularity of HTTPs. However, progressive download does not support dynamic bitrate adaptation efficiently leading to poor performance especially in highly changing network conditions. The history of Internet-based video streaming is illustrated in Figure 4.1.

4.2 HTTP Adaptive Streaming (HAS) Framework

Dynamic Adaptive Streaming over HTTP (MPEG-DASH) [2] is an adaptive HTTP-based protocol designed for streaming media over the Internet. Conventional HTTP web servers and an Adaptive-Bitrate Streaming (ABS) mechanism are used to enable high-quality video content streaming over the Internet. MPEG-DASH is the commonly used video adaptive-bitrate HTTP-based streaming approach that has been adopted by streaming technologies such as Adobe HTTP Dynamic Streaming (Adobe HDS), Apple HTTP Live Streaming (Apple HLS), and Microsoft Smooth Streaming. An adaptation mechanism during a video streaming session at the client decides the video bitrate that have to be requested for each video chunk or segment. MPEG-DASH is an adaptive HTTP-based protocol designed for streaming media over the Internet. The technology is used to transport segments of live and on-demand video content from web servers to viewers' devices [3].

An adaptation mechanism during a video streaming session at the client decides the video bitrate that has to be requested for each video chunk or segment. It is important to mention that HAS reduces the interruption rate by allowing a client to adapt the video bitrate to the throughput. That way HAS enables a high video quality to be delivered to the DASH client's device compared to progressive download streaming approach discussed above. Following the release and first launch of HAS in 2007, the 3GPP in collaboration with the Universal Mobile Telecommunications Systems – Long Term Evolution (UMTS-LTE) released the HAS standard in 2009. As part of their collaboration with MPEG, 3GPP Release 10 adopted DASH with the inclusion of specific codecs such as H.265, H.264, and VP9 for use over wireless networks. The MPEG-DASH second edition was released in 2014. Apart from standardization bodies, the DASH Industry Forum (DASH-IF) is another body that has been promoting the adoption of MPEG-DASH since 2012. The DASH-IF is made up of major streaming and media companies, including Netflix, Microsoft, Ericsson, Google, Samsung, Adobe, and many others. The DASH-IF is also responsible for creating guidelines on the usage of DASH for different use cases in practice.

The general DASH architecture is shown in Figure 4.2. It consists of the DASH server and DASH clients and the delivery network. On the server-side, videos are encoded into multiple levels. These levels are usually identified with the video bitrates which are generated by compressing/encoding the raw video with

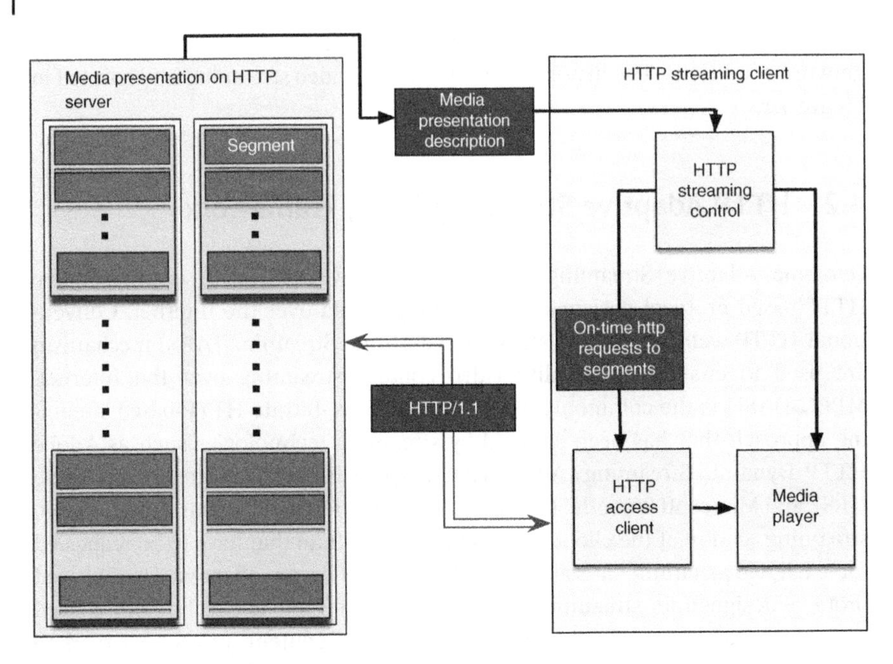

Figure 4.2 DASH architecture.

one of the standard codes (i.e. H.264, H.265/HEVC, H. 266 (Versatile Video Coding (VVC), VP9, etc.). The video bitrate is directly proportional to video quality. However, the relation between the video quality and its bitrate is not always linear. Two main approaches namely the constant bitrate (CBR) and the variable bitrate (VBR) are normally used for encoding the video bitrate. With the CBR encoding schema, the video is compressed with a constant level of bitrate and variant quality levels along the video, while the other approach keeps the quality constant at the cost of bitrate variability. The videos on the server-side are segmented into small chunks of the same video duration (i.e. 2–12 s) using GPAC MP4Box [4]. To provide a seamless streaming and smooth switching between the different bitrate, each chunk is encoded independently. Therefore, in most cases, each segment starts with intracoded frame (I-frame), so it allows the client to start playing that segment once it been downloaded.

As shown in Figure 4.3, each DASH video on the server-side is coupled with its metadata file that is denoted as Media Presentation Description (MPD). MDP is an Extensible Mark-up Language (XML) file that hosts all the content-related information (i.e. video bitrates, resolutions, URL location of the video chunks, video codec). Furthermore, this information is structured heretically within the MPD file with the period as the root element of the XML file. Each period represents video object with specific duration and it hosts one or more adaptation

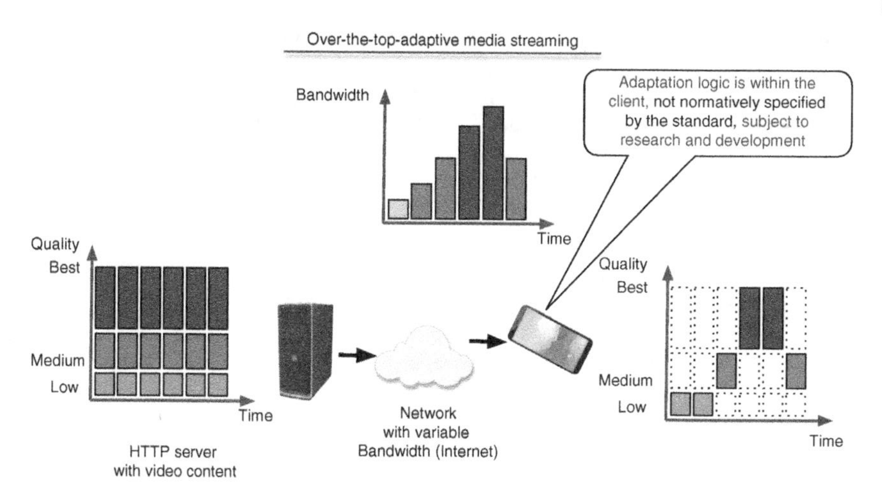

Figure 4.3 An overview of DASH players video streaming QoE.

sets that contain all versions of the content each known as a representation. Each representation hosts the initialization segment that contains metadata of such representation along with other video segments. These segments have different quality (low, medium, and high) based on the video encoded bitrates (e.g. 360p, 480p, 720p). The client side receives a video resolution based on the network conditions available. For example when the network condition is good, a high video quality is streamed at the client's side (Table 4.1).

Figure 4.4 depicts the behavior of the DASH system for delivering the video content to clients. As shown in Figure 4.4, each DASH client starts the video by requesting the MDP file from the server side. Upon the reception of the MPD file, the client issues a set of HTTP requests (typically in chronological order) for the video segments [10]. All the downloaded segments are stored in the video playback buffer (buffering phase) before the video starts playing (playing phase). Further, each playback buffer (video buffer) has a maximum size (i.e. mostly 30 s). Therefore, it is often that the client has an ON/OFF download patterns, especially when the arrival time for requested segment is less than the playtime [3]. To provide a smooth playback streaming, select the most appropriate segment on the right time, each DASH player is equipped with a local Adaptive Bitrate Algorithm (ABR) that adapts the quality and requests the bitrate level that matches the network conditions. Different approaches of the ABR algorithms have been presented in the literature [3] including throughput prediction mechanisms. Most of client-side bitrate adaptation approaches employ the segment download time and/or playback buffer occupancy level to estimate the video quality based on the network conditions. Most of the bitrate adaptation algorithms have been built to run under

Table 4.1 A comparison of HTTP adaptive streaming solutions.

HAS category	Company	Video codec	Segment length (s)	Data description	Format
Microsoft Smooth Streaming [5]	Microsoft Corporation	H.264, VC-1	2	Manifest (XML)	fMP4
Apple HTTP Live Streaming (HLS) [6]	Apple Inc.	H.264	10	Playlist file (M3U8)	M2TS, *.ts files
Adobe HTTP Dynamic Streaming (HDS) [7]	Adobe Systems Inc.	H.264, VP6	2 - 5	Manifest (F4M)	fMP4
MPEG-DASH [8]	Standard	Any	Not specified	Media Presentation Description (MPD) files (XML)	MP4 or M2TS
3GP-DASH [9]	Standard	H.264	Not specified	MPD files (XML)	3GPP File Format

M2TS, MPEG-2 Transport Stream; fMP4 is a fragmented MP4; MPD = Media Presentation Description

HTTP/1.1, whereas every chuck is delivered under the HTTP request/response pattern. However, HTTP/2 [11] has been deployed recently to act as the application protocol within DASH architecture. HTTP/2 has the potential for enhancing the delivery of DASH traffic, especially with live streaming. DASH traffic can be easily traversed through NATs and firewalls using HTTP/2, and content providers can utilize the conventional Web servers for hosting and cashing DASH contents. Figure 4.4 illustrates the concept of DASH assuming a video rate adaptation method (e.g. throughput-based, buffer-based, or hybrid) for video streaming. It can be observed that the client, based on its network condition, adapts the quality of the video to provide a smooth streaming experience to the end user.

HAS's success can be due to the following advantages over standard streaming technologies: (i) video service providers can modify video bitrates to deliver multiple quality levels to meet end-user requests; and (ii) clients can be offered several QoE-tailored customized service levels and/or pricing schemes [12]. It is worth mentioning that DASH allows for both live and on-demand media stream transmission over HTTP. Optimized encoding technologies such as multipass and VBR encoding can help on-demand video streaming apps. Netflix, for example

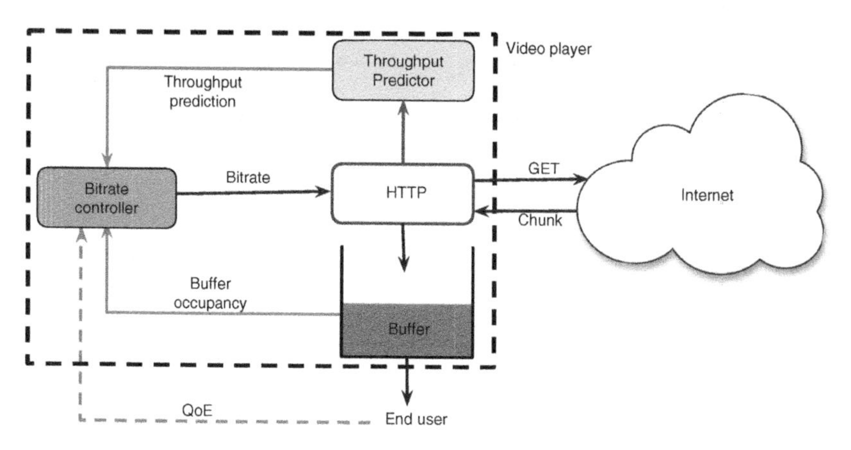

Figure 4.4 Example of client-centric HTTP adaptive streaming session.

suggested a content-aware video encoding optimization [103] technique in which several ideal resolution-bitrate pairs are calculated using content information, resulting in improved service quality [3].

Despite the benefits offered by HAS because of its decentralized nature, there are still certain limitations, particularly when numerous DASH clients compete for shared network resources. The HAS challenges includes (i) bitrate switching that leads to instability [13]; (ii) network resource underutilization; and (iii) QoE unfairness are some of the HAS-related concerns [14]. These issues continue to be a major problem for mobile network operators and video content providers, and they are exacerbated in heterogeneous environments such as the stadium or shopping mall contexts. Based on these considerations, MPEG-SAND standard was proposed to improve the distribution of DASH material through the use of centralized nodes within the network [15, 16].

4.3 Server and Network-Assisted DASH (SAND)

As an extension to MPEG-DASH standard, the Server and Network-assisted DASH (SAND) was developed so as to facilitate and improve the efficiency of video streaming services; SAND provides messages between various network elements and DASH clients. This is achieved by providing information about real-time operational characteristics of servers, caches, networks, proxies, and DASH clients' streaming status and performance [17, 18].

The MPEG-SAND architecture (Figure 4.5 enables cooperation between the operations of media server and DASH client during a video streaming session. It also provides the standardized interfaces toward realizing the (i) video streaming

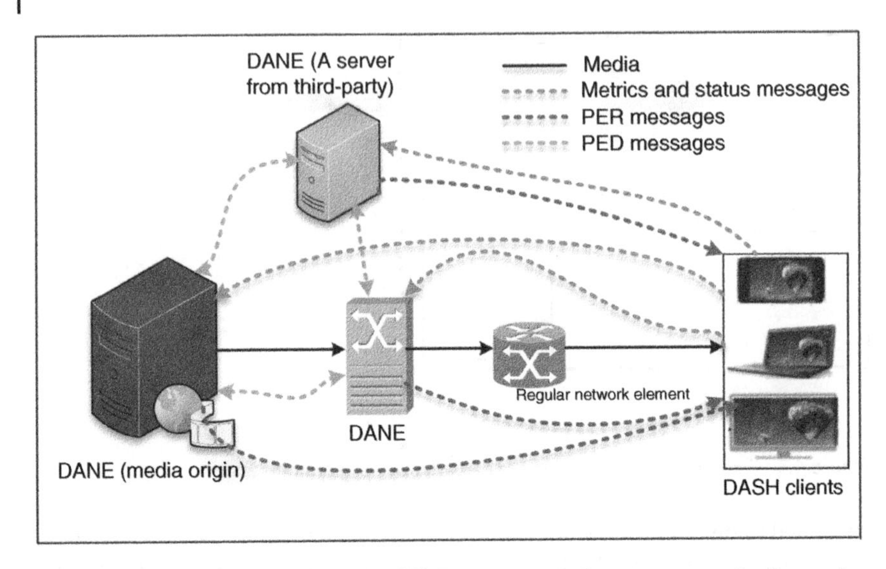

Figure 4.5 MPEG-SAND architecture. DANEs can communicate among each other and with the DASH clients to optimize the end-to-end video quality delivery. Source: Barakabitze et al. [3] / with permission of IEEE.

improvements using intelligent processing, caching, and QoE-delivery optimization mechanisms on the network and/or server side using feedback from the DASH clients on the requested segments and network bandwidth; (ii) improved QoE adaptation on the DASH client side based on the server/network-side information (e.g. network throughput, QoS/QoE, video segments) [19]. The SAND architecture shown in Figure 4.5 is designed for supporting different MPEG-DASH standards such as MPEG-I [20] for immersive media and MPEG-V [21] information representations to enable interoperability between virtual worlds (e.g. digital content provider of a virtual world, gaming, and simulation), and between real and virtual worlds (e.g. sensors, actuators, vision and rendering, and robotics). SAND has been designed to facilitate content awareness through the monitoring of DASH-based services using peer-to-peer or multipoint communication with or without management between servers and DASH clients.

The SAND architecture is made up of three types of components namely DASH-aware Network Elements, DASH clients, and DASH-Aware Network Elements (DANE). SAND includes also normal network elements (that aren't aware of DASH) that are present on the path between the origin server and the DASH clients (e.g. transparent caches). DANE communicates with DASH clients while possessing just rudimentary knowledge of the protocol. DANE consists of knowledge and minimum intelligence about DASH clients, servers, multi-access edge computing (MEC) caches, and content delivery networks. For example, DANE nodes may be aware that the provided objects, such as the mpd or DASH segments, are DASH-formatted. They can then prioritize, parse, and even alter such items in this manner. Three messages are defined in the MPEGAV-SAND architecture, namely (i) the Parameters Enhancing Reception (PER); (ii) Parameters Enhancing Delivery (PED); and (iii) metrics and status messages. The PER messages are sent from DANEs to DASH clients so as to enhance and improve the video streaming quality adaptation, while PED messages are exchanged between DASH clients and DANES. The metrics and status messages are sent from DASH clients to DANES [3]. Using SAND status messages, DANE nodes learn about the state of DASH clients. QoE metrics given to the network by DASH clients, for example can be utilized for monitoring and simplifying video data rate prioritization and optimization strategies. PER messages can be used by a DANE node to inform DASH clients about the available network bandwidth. They can also inform clients about segments that have already been cached by the DANE, allowing clients to request these segments based on their device capabilities [3, 22].

A PED message can also be used by the server to transmit information about the streamed video to a specific network delivery element/node. It is worth noting that a DANE element can be also a third-party server that receives metrics messages from DASH clients and provides SAND messages to the clients. The MPEG-SAND messages are sent through HTTP in XML format and must adhere to a strict syntax prescribed by the standard [17]. The MPEG-SAND standard, in comparison to client- or server-based solutions, is a critical enabler for solutions that require collaboration between service/application and network providers. SDN can be used to provide a centralized control element for a practical and scalable deployment of the SAND method. SDN and NFV-based QoE-driven applications and QoE-aware network management strategies look to be critical solutions for ensuring end users' QoE at a low cost. Clayman et al. [23] propose a SAND video streaming architecture based on SDN and NFV, in which virtual cache instances are built on-demand based on live DASH properties and characteristics. Guillen et al. [24] propose SAND/3, an SDN-based QoE control approach for DASH over HTTP/3 that combines information from various layers which are orchestrated by the controller to select the best QoE for the end users.

The SAND/3 architecture as indicated in Figure 4.6 combines user, device, service, and network-level information. It consists of three components: user, network, and application modules. This module gathers the end-user identity and associated devices from a third-party service to which the user has subscribed (for example Netflix) and stores them in a user profile repository [24]. The network module performs TE by routing packets through the most appropriate paths with the best QoS. The network statuses are collected from all network elements and monitored by the network monitor submodule. The QoE manager uses the current state of the network, user profile, and the specific service policies to recommend

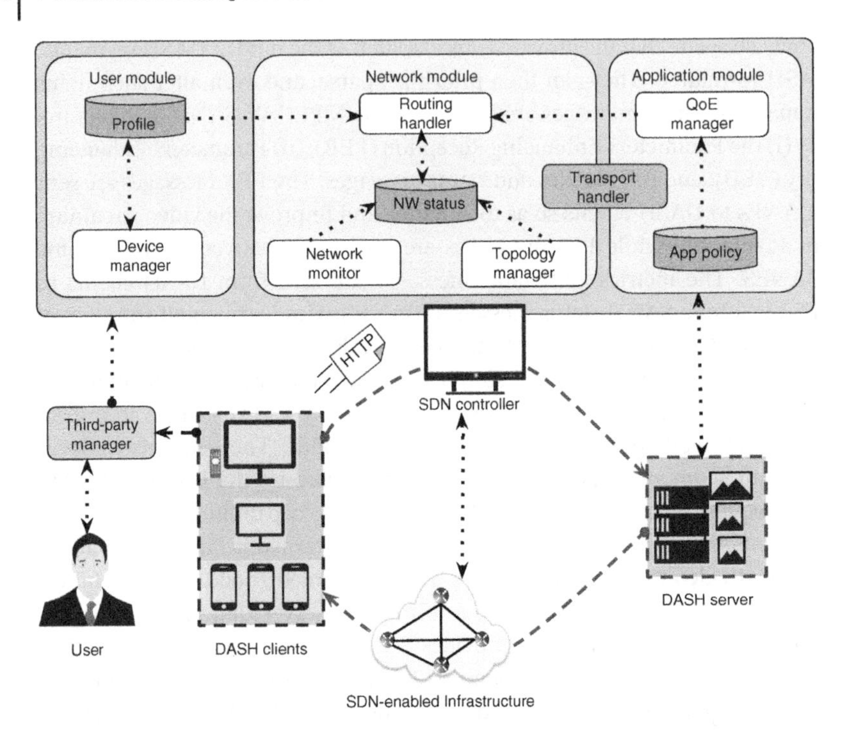

Figure 4.6 SAND/3 architecture.

the most suitable settings for the transmission, which is handled by the transport handler submodule. SAND/3 can minimize video interruptions, reduce the number of quality shifts in at least 40%, and increase the amount video contents downloaded by at least 20% compared to the current client adaption implementations such as throughput-based, buffer-based, and segment-aware rate algorithms.

4.4 Multimedia Delivery Chain and Service Management Issues

The OTTPs, ISPs, CDNs, transit providers, and IXPs are all part of the multimedia service delivery chain, with each player having a particular role based on their mutual agreements. On the OTTP side, multimedia content distribution operations can be classified into two categories: (i) content distribution and replication; (ii) client request redirection. The TE abstractions, which include optimal routing techniques, prioritization, and dynamic resource allocation, make up the majority of network management operations on the ISP side. There are two techniques to distribute and replicate content from the source server to CDNs and Surrogate

Servers (SSs): (i) hierarchical and (ii) flat. Content in the hierarchical organization is distributed/replicated in a multitree fashion with HTTP redirection used to move content up and down the tree between the source server, CDNs, and SSs. To avoid cache misses, information about the content's location might be integrated into the overlay in this situation. The drawback of hierarchical distribution is that cache misses and higher end-to-end delay can be encountered when the video content is not available in the last mile SS.

The request redirection mechanism can be based either on Domain Name System (DNS) unicast or on a combination of DNS and anycast for assignment of the nearest server to the clients. DNS-based request redirection relies solely on link-layer information to assign CDN/SS to customers, making it unable to respond to congestion in the ISP's network. In this instance, the OTTP is unable to provide enough video quality to end users, while the ISP is forced to transmit data over a less-than-optimal route because no information is communicated between the OTTP and ISPs. On the worst-case scenario, the client is assigned to a CDN outside of the ISP's network, which not only reduces quality but also increases the ISP's costs and increases network load in the main network. In this instance, SDN provides a chance for ISPs to aid OTTP in making the best CDN/SS choice for their network [25].

The collaboration between ISP and CDN can provide them with triple-win benefits. For example, ISP can gain better QoE-traffic management and efficient network resources utilization leading to a reduced cost of operation and investment. The CDN can acquire the network information and use them to improve the performance of link load and user's experience based on location [26]. An SDN-based CDN–ISP collaboration solution can greatly improves the efficiency of content delivery and the end-users QoE. The CDN–ISP collaboration architectures in this direction are proposed in [27–29]. Akyildiz et al. [27] propose SoftAir, a software-defined architecture for 5G wireless networks that improve the efficiency of content delivery. The architecture adopts different approaches such as distributed and collaborative traffic classifiers and mobility-aware control traffic balancing to optimize the control of traffic control and network monitoring. Wichtlhuber et al. [29] proposed an SDN-based architecture that provides a fine grained, integrated traffic engineering for CDN traffic in the ISPs network. The CDN provider is given an ability to decide on the selection of surrogate servers. Wang et al. [28] propose NetSoft, a software-defined decentralized mobile network architecture for 5G to improve the efficiency of content delivery in 5G systems.

The QoE-aware OTT-ISP collaboration in service management has been investigated in recent years for improving QoE of users. Floris et al. [30] propose a QoE-aware service management reference architecture for a possible collaboration and information exchange among OTT and ISP in terms of technical and economic aspects. A joint venture, customer lifetime value, and QoE-fairness mechanisms are proposed to maximize the revenue by providing better QoE to customers paying more. Authors in [31] propose a PPNet, a framework that enables isolation between the SP's web and video provider's interfaces. PPNet also allows CDN-ISP collaboration while preventing the ISP's access to the video request and availability information. Ahmad et al. [32] propose a novel service delivery approach that is purely driven by the end user's QoE, while considering the collaboration between OTT and ISP. Recent developments of network slicing for multidomain orchestration and management provide a realization of E2E management and orchestration of resources in 5G and beyond sliced networks [33]. The collaboration between verticals (ISPs, CDNs, OTTs, etc.,) in future softwarized networks are required that ensures multimedia streaming service requests from different domains that are mapped into multioperator and multiprovider technology domains, while matching each service ELA/SLA requirements [33].

4.5 Conclusion

Multimedia streaming services by OTTP are using HTTP adaptive video streaming based on the MPEG-DASH standard. While the MPEG-DASH standard can deliver outstanding video quality to end users, there are several challenges when multiple clients compete for the same restricted resources, resulting in QoE inequity, video instability, and network resource underutilization. To address these issues, the MPEG-SAND standard was recently suggested with the goal of improving DASH content delivery through its capacity to give real-time network, server, and streaming DASH client information. DASH clients can advise DANEs about the operating video data rate, a set of segments, and video quality they require. DASH clients can get information from DANEs about the video segment, network speed availability, and segment caching status. However, OTTPs and ISPs may collaborate in a collaborative QoE service management paradigm employing (i) client request, redirection, and optimal route; and (ii) content distribution and replication from the standpoint of the multimedia delivery chain and service management.

Bibliography

1 Majumdar, A., Sachs, D.G., Kozintsev, I.V. et al. (2011). Multicast and unicast real-time video streaming over wireless LANs. *IEEE Transactions on Circuits and Systems for Video Technology* 12 (6): 524–534.

2 Sodagar, I. (2011). The MPEG-DASH standard for multimedia streaming over the internet. *IEEE MultiMedia* 18 (4): 62–67. https://doi.org/10.1109/MMUL.2011.71.

3 Barakabitze, A.A., Barman, N., Ahmad, A. et al. (2019). QoE management of multimedia services in future networks: a tutorial and survey. *IEEE Communication Surveys and Tutorials* 22.

4 GPAC: Multimedia Open Source Project. https://gpac.wp.imt.fr/mp4box/ (accessed 11 August 2022).

5 Microsoft (2017). Microsoft Silver light Smooth Streaming. https://www.microsoft.com/silverlight/smoothstreaming/ (accessed 06 March 2018).

6 Apple (2017). Apple HTTP Live Streaming. https://developer.apple.com/streaming/ (accessed 06 March 2018).

7 Adobe (2017). Adobe HTTP Dynamic Streaming (HDS). https://www.adobe.com/devnet/hds.html (accessed 06 March 2018).

8 ISO/IEC 23009-1:2014 (2017). *Preview Information technology – Dynamic adaptive streaming over HTTP (DASH) – Part 1: Media presentation description and segment formats.* https://www.iso.org/standard/65274.html (accessed 06 March 2018).

9 ETSI TS 126 247 V14.1.0 (2017). *Universal Mobile Telecommunications System (UMTS); LTE; Transparent end-to-end Packet-switched Streaming Service (PSS); Progressive Download and Dynamic Adaptive Streaming over HTTP (3GP-DASH) (3GPP TS 26.247 version 14.1.0 Release 14).* http://www.etsi.org/deliver/etsi.ts/126200.126299/126247/14.01.00.60/ts.126247v140100p.pdf/ (accessed 10 September 2019).

10 Huang, T., Zhang, R.-X., Zhou, C., and Sun, L. (2018). QARC: Video quality aware rate control for real-time video streaming based on deep reinforcement learning. *Proceedings of the 26th ACM International Conference on Multimedia*, October 2018.

11 Kua, J., Armitage, G., and Branch, P. (2017). A survey of rate adaptation techniques for dynamic adaptive streaming over HTTP. *IEEE Communication Surveys and Tutorials* 19 (3): 1842–1866.

12 Seufert, M., Egger, S., Slanina, M. et al. (2015). A survey on quality of experience of HTTP adaptive streaming. *IEEE Communication Surveys and Tutorials* 17 (1): 469–492.

13 Rodríguez, D.Z., Wang, Z., Rosa, R.L., and Bressan, G. (2014). The impact of video-quality-level switching on user quality of experience in dynamic adaptive streaming over HTTP. *EURASIP Journal on Wireless Communications and Networking* 216 (1). 1–15.

14 Bentaleb, A., Begen, A.C., Zimmermann, R., and Harous, S. (2017). SDNHAS: An SDN-enabled architecture to optimize QoE in HTTP adaptive streaming. *IEEE Transactions on Multimedia* 19 (10): 2136–2151.

15 Cofano, G., De Cicco, L., Zinner, T. et al. (2017). Design and experimental evaluation of network-assisted strategies for HTTP adaptive streaming. *ACM*

Transactions on Multimedia Computing, Communications, and Applications 35 (11): 1–24.

16 Bhat, D., Rizk, A., Zink, M., and Steinmetz, R. (2017). Network assisted content distribution for adaptive bitrate video streaming. *Proceedings of the 8th ACM on Multimedia Systems Conference*, pp. 62–75. June 2017.

17 ISO/IEC 23009-5 (2017). *Information Technology - Dynamic Adaptive Streaming over HTTP (DASH) - Part 5: Server and Network Assisted DASH (SAND). Standard.*

18 Thomas, E., van Deventer, M.O., Stockhammer, T. et al. (2017). Enhancing MPEG DASH performance via server and network assistance. *SMPTE Motion Imaging Journal* 126 (1): 22–27.

19 DASH Industry Forum (2018). Guidelines for Implementation: DASH-IF SAND Interoperability. https://dashif.org/docs/DASH-IF-SAND-IOP-v1.0.pdf// (accessed 22 June 2021).

20 MPEG-I: ISO/IEC 23090. *Coded Representation of Immersive Media*. https://mpeg.chiariglione.org/standards/mpeg-i// (accessed 22 June 2021).

21 MPEG-V: ISO/IEC 23005 *Media Context and Control*. https://mpeg.chiariglione.org/standards/mpeg-v// (accessed 22 June 2021).

22 Bentaleb, Y.A., Begen, A.C., and Zimmermann, R. (2018). QoE-aware bandwidth broker for HTTP adaptive streaming flows in an SDN-enabled HFC network. *IEEE Transactions on Broadcasting* (99): 1–15.

23 Clayman, S., Kalan, R.S., and Sayit, M. (2018). Virtualized cache placement in an SDN/NFV assisted SAND architecture. *IEEE International Black Sea Conference on Communications and Networking (BlackSeaCom).*

24 Guillen, L., Izumi, S., Abe, T., and Suganuma, T. (2019). SAND/3: SDN-assisted novel QoE control method for dynamic adaptive streaming over HTTP/3. *Electronics* 8 (8): 864.

25 Nygren, E., Sitaraman, R.K., and Sun, J. (2010). The Akamai network: a platform for high-performance internet applications. *ACM SIGOPS Operating Systems Review* 44 (3): 2–19.

26 Qingmin, J., Renchao, X., Tao, H., Jiang, L., Yunjie, L. (2017). The collaboration for content delivery and network infrastructures: a survey. *IEEE ACCESS* 5: 18088–18106.

27 Akyildiz, I.F. et al. (2015). SoftAir: A software defined networking architecture for 5G wireless systems. *IEEE Network* 85: 1–18.

28 Wang, H. et al. (2015). SoftNet: A software defined decentralized mobile network architecture toward 5G. *IEEE Network* 29 (2): 16–22.

29 Wichtlhuber, M. et al. (2015). An SDN-based CDN/ISP collaboration architecture for managing high-volume flows. *IEEE Transactions on Network and Service Management* 12 (1): 48–60.

30 Floris, A. et al. (2018). QoE-aware OTT-ISP collaboration in service management: architecture and approaches. *ACM Transactions on Multimedia Computing, Communications, and Applications* 14 (2): 1–24.

31 Akpina, K. and Hua, K.A. (2020). PPNet: Privacy protected CDN-ISP collaboration for QoS-aware multi-CDN adaptive video streaming. *ACM Transactions on Multimedia Computing, Communications, and Applications* 16 (2): 1–23.

32 Ahmad, A. et al. (2018). QoE-centric service delivery: a collaborative approach among OTTs and ISPs. *Computer Networks* 110: 168–179.

33 Barakabitze, A.A., Arslan, A., Rashid, M., and Hines, A. (2020). 5G network slicing using SDN and NFV: a survey of taxonomy, architectures and future challenges. *Computer Networks* 167: 1–40.

5

QoE Management of Multimedia Services Using Machine Learning in SDN/NFV 5G Networks

This chapter summarizes the state-of-the-art approaches to Quality of Experience (QoE) Management using Software-Defined Network (SDN) and Network Function Virtualization (NFV) in future networks. It highlights a variety of management systems and architectures as well as benchmarking studies that have been carried out. It gives the reader a taste of the concepts and approaches that are being proposed and adopted to deliver QoE in future networks.

QoE management of multimedia services has continued to have a major and increasing impact on the quality of videos delivered to the end users. Due to the increasing popularity and use of video streaming services (e.g. YouTube and Netflix) on smart devices such as smartphones, QoE is now at the core of Internet Service Providers (ISPs), mobile operators, and Over-The-Top (OTT) service providers. Delivering high video quality to the end users is very important for the continued success of such services. The need and increased user demand of services with excellent quality have triggered ISPs, OTT providers, and telecom operators to upgrade their systems for managing multimedia services. This has made them to invest in emerging network softwarization and virtualization technologies such as SDN, NFV, Multi-access Edge Computing (MEC), and Cloud/Fog Computing. The mounting pressure from new emerging multimedia services (e.g. 4K/8K/12K and 360° video streaming applications) and various use cases in 5G networks such as Machine-Type Communication (MTC), network-controlled Device-to-Device (D2D) communications, and Massive Internet of Things (MIoTs) is driving this transformation and the overall upgrading of their operation and management systems. Academia and industry have made significant efforts to optimize the end-user's QoE since the standardization of the Dynamic Adaptive Streaming over HTTP (MPEG-DASH)[1] and Server and Network-assisted DASH (SAND) [1]. SAND improves the delivery of DASH content with excellent quality to users by using well-defined messages that are

1 https://mpeg.chiariglione.org/standards/mpeg-dash.

Multimedia Streaming in SDN/NFV and 5G Networks: Machine Learning for Managing Big Data Streaming, First Edition. Alcardo Barakabitze and Andrew Hines.

exchanged between DASH client and network elements (e.g. SDN controller). The SAND messages provide information in real time about the streaming characteristics of video servers, Content Delivery Networks (CDNs), caches, networks, and status of DASH clients' performance. Today, different approaches for improving the end-user's QoE using SDN and/or NFV have been proposed from the academia and industry [2]. This chapter provides QoE management strategies with a focus on the following areas: (i) QoE-centric routing, (ii) QoE optimization using MultiPath Transport Control Protocol (MPTCP)/Segment Routing (SR) and Quick UDP Internet Connections (QUIC) approaches, (iii) Server and network-assisted for QoE optimization, (iv) QoE-centric fairness and personalized QoE control, and (v) QoE-driven optimization for video streaming services using Machine Learning (ML).

5.1 QoE-Centric Routing Mechanisms for Improving Video Quality

Much research focuses on traffic rerouting to improve the received quality at the end user. Egilmez et al. [3] present OpenQoS, an optimization framework that can forward decisions at the control layer of OpenFlow-based networks to enable dynamic Quality of Services (QoS) for the end users. The aim of OpenQoS is to enable QoS-enabled scalable encoded video streaming while guaranteeing the QoS levels. The enhancement layers of scalable video are treated as best-effort flows or level-2 QoS, while the based layer is treated as a level-1 QoS flow. Cetinkaya and coworkers [4] propose a segment-based routing strategy using SDN while taking into account the heterogeneous DASH clients. The video bitrate, throughput, the segment, and path length are the parameters considered to select the best path for routing the video flows from the server to the DASH client. Dobrijevic et al. [5] proposed a QoE-centric flow-routing mechanism that utilizes QoE estimation models in order to maximize the end-user's QoE for video streaming services. Bouten et al. [6] employ a probability-based search mechanism to propose a dynamic server selection algorithm that is able to detect and select the optimal server based on its network characteristic to deliver the video to clients with an improved quality. The end-user's QoE is achieved based on the collected knowledge on the network characteristics, however, without requiring any adaptations overhead or active measurements to the server.

Al-Jawad et al. [7] propose LearnQoS, an approach that employs Reinforcement Learning (RL) and Policy-based Network Management (PBNM) for QoS/QoE optimization using SDN. LearnQoS utilizes the PBNM in order to facilitate the traffic engineering management and QoS/QoE policy violation over SDN during multimedia delivery to the end user. A QoE-based routing strategy of video streaming services for overlay and ISP networks is proposed by Calvigioni et al. [8]. Authors develop a QoS–QoE model that can incorporate different QoE metrics (e.g. rebufferings) and predict their impact on different video adaptation logic and DASH clients' resolution in changing network conditions. Huang et al. [9] use a Deep Reinforcement Learning (DRL) to propose an adaptive multimedia traffic control strategy over SDN. The proposed approach can directly learn to control multimedia traffic by allocating an appropriate path with unequivocal bandwidth for each video flow. Khalid et al. [10] introduce a Device-Aware Network-assisted Optimal Streaming service (DANOS) that can achieve an inter-domain adaptive bitrate streaming in SDN. DANOS can achieve a maximized video quality for the end users while considering their subscription levels, device capabilities, and network constraints. Moreover, a QoE optimization model that can determine the best streaming path of the video packets of DASH clients is proposed in [11]. The authors introduce an SDN-based fair service differentiation architecture that considers the characteristics of DASH applications. The proposed mechanism can provide QoE-fairness among DASH clients that belong to the same service class. [12] demonstrates a Scalable QoE-Aware Path Selection (SQAPE) scheme for large-scale SDN-based mobile networks. Using a centralized control strategy of SDN, SQAPE provides fine-grained control for per-user QoE-aware path deployments across the network. As shown in Figure 5.1, it consists of the *QoS measurement, and QoE predictor* components, which are decoupled from the SDN controller.

As shown in Figure 5.1, steps (1)–(4) involve QoS measurements to determine the performance of video streaming. After every 60 seconds, the network monitoring is performed to measure the QoS based on packet loss, delay, and bandwidth of the link. Such metrics are selected based on previous investigations of QoS to QoE mapping [13], where packet loss and delay metrics were also found to affect the video quality in real-time video streaming services. SQAPE in steps (5)–(8) relies on the QoS-to-QoE mapping function to compute per-path Mean Opinion Score (MOS) estimation. The estimated MOS, along with a link utilization, is then used to compose the delay metric, which in turn determines optimized QoE-aware paths to be deployed in SDN-enabled networks. Venkataraman et al. [14] investigate the effects of the Internet path selection on video QoE using large-scale measurements from 62 vantage points on PlanetLab.[2] 62% of paths to broadband hosts and up to 89% of paths to servers are recovered using alternative paths. The results indicate that alternate paths are very beneficial to improve the end-user's QoE for online video streaming services. The authors in [15] formulate the QoE-based path selection strategy using a mixed-integer linear programming in multihop wireless ad hoc networks to enhance resource utilization and the users' QoE.

2 http://www.planet-lab.org.

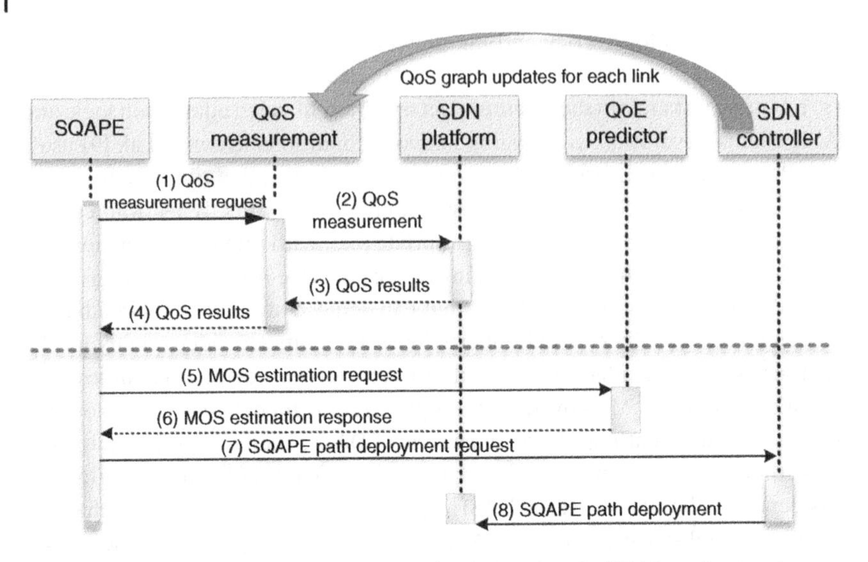

Figure 5.1 Scalable QoE-aware path selection in large-scale SDN-based networks.

A Pseudo-Subjective Quality Assessment (PSQA) tool and a multi-commodity flow-routing model are employed to describe flows in wireless ad hoc networks and maximize the overall MOS under link and time constraints. The performance of a Better Approach To Mobile Ad hoc Networking (BATMAN) is evaluated in [16] for Voice over IP (VoIP) services on low-battery consumption over Mobile Ad Hoc Networks (MANETs). The BATMAN performance for VoIP services is also assessed using the MOS score and QoS parameters (e.g. delay, packet loss) to investigate the effect of the number a density of ad hoc nodes. The evaluation results indicate that BATMAN implementation is suitable enough for managing MANETs formed by low-energy consumption nodes as well as supporting the VoIP services.

5.2 MPTCP/SR and QUIC Approaches for QoE Optimization

MPTCP has been mostly been used to improve the aggregate system throughput and QoE of the video streaming client. Barakabitze et al. [17] present Traffic Engineering (TE) management solutions in SDN by employing both MPTCP and SR to facilitate an efficient routing of multimedia services between end-points. The authors propose to use multiple shortest paths for MPTCP subflows transmission in SDN/NFV networks. Multi-flow commodity and Shortest Path Model (MCSPM) models are applied to select important intermediate nodes to perform

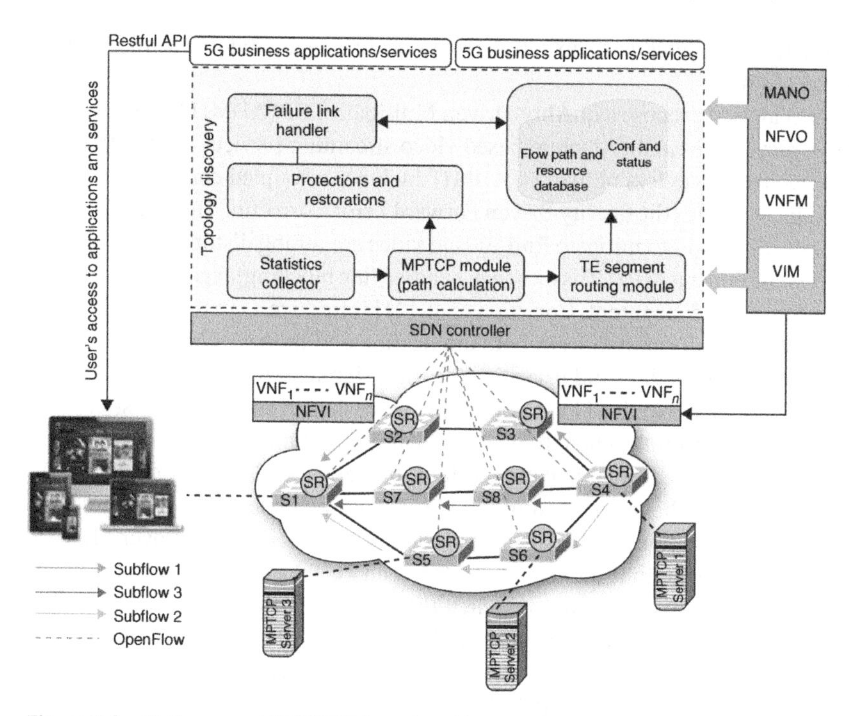

Figure 5.2 QoE -aware MPTCP/SR-based architecture in softwarized 5G networks.

source routing using SR paradigm. Barakabitze et al. [18] further propose a novel QoE-aware, SDN-based MPTCP/SR framework for 5G networks as shown in Figure 5.2. The proposed approach can control dynamically the number of MPTCP subflows and efficiently use the available network resources by forwarding the network traffic through multiple disjointed paths. The MPTCP-based softwarized architecture consists of the following modules: QoE management module, failure recovery handler, database, TE-SR, MPTCP-based flow manager, and the Network Information Collector (NIC). The TE-SR Module maps the computed MPTCP subflow paths to SR paths which in turn generates a list of segment labels that are used for video transmission. The QoE management module is responsible for QoE-driven network management and service management for enhancing the DASH clients' QoE [17, 18]. It also allocates the available network resources and performs video quality optimization and monitoring (network and user feedback) of video streaming services. The database stores the network monitoring status, reports, and various configuration parameters. It also keeps and maintains the SR subflow paths that have been used with their QoE requirements for future. The failure recovery handler module detects the link/node failures in softwarized networks. It also provides multipath protections and dynamic link recovery to ensure that any failed parts in softwarized network are configured within a short period of time.

Wu et al. [19] propose a quAlity-Driven MultIpath TCP (ADMIT) approach that can achieve high-quality mobile-based video streaming services with MPCTP in heterogeneous wireless networks. ADMIT includes a coupled congestion control and it incorporates the quality-driven Forward Error Correction (FEC) [20] coding and rate allocation to mitigate End-to-End video streaming distortion. In order to address the current MPTCP issues (e.g. head-of-line blocking) regarding the transport of multimedia content, Corbillon et al. [21] introduce a cross-layer scheduler that leverages information from both application and transport layers to re-order the transmission of data and prioritize the most significant parts of the video. The viewers' QoE is maximized by decoding the video data even in difficult streaming conditions, for example when there is a small buffer and the video bitrate is approximately close to the available bandwidth in the network. The question regarding on how MPTCP is beneficial to mobile video streaming is answered by James et al. [22] who analyze the performance of various scenarios for DASH over MPTCP. The results indicate that the end-users' QoE can be achieved with MPTCP when the primary path is very high even when there is high variability of bandwidth in the network. Authors also indicate that MPTCP can benefit when two paths with stable bandwidths are used. MPTCP can also benefit when the secondary path has small bandwidth. Again, MPTCP can also benefit higher video bitrate with no impact on quality switching and startup delays when both paths have stable and ample bandwidth.

Han et al. [23] propose an MP-DASH, a novel multipath strategy that can enhance MPTCP to support adaptive video streaming under user-specified interface preferences. MP-DASH consists of two main components, namely the MP-DASH scheduler and the video adapter. The scheduler receives video segments and the user's preferences to determine the best fetching strategy of the video segments over multipath. The video adapter is a lightweight component that handles the interaction between the DASH rate adaptation algorithms and the MP-DASH scheduler. The performance results of MP-DASH indicate that the cellular usage and the radio energy consumption can be reduced by 99% and 85%, respectively, with a negligible QoE degradation compared to the native MPTCP approach. Nam et al. [24] achieve faster download speed and an improved QoE, where SDN controller is used to add or remove MPTCP transmission paths in order to reduce large number of out-of-order packets which may cause poor performance and degrade the end-users' QoE. Mohan et al. [25] conduct measurements-driven study with regard to multipath in mobile devices to investigate the path utilization, Transport Control protocol (TCP) throughput, handshake latency, video streaming bitrate, and web page load time. The authors conduct passive measurements with crowd-sourced probing on participants' phones to capture the end-user's QoE over multipath under changing network conditions.

Mohan et al. [25] perform measurement study of MPTCP over multi-carrier Long Term Evolution (LTE) connections in normal daily working mobility scenarios of mobile services for a period of five months. The authors collect data for a period of five months in controlled environment and in the wild. The main objective is to understand the impact of MPTCP with regard to mobility, video application workload, and last-mile video quality. Hayes et al. [26] employ MPTCP

Table 5.1 A summary of QoE-centric management strategies of multimedia streaming services in future networks.

Strategy	Major contributions	Objectives/Functionality
Server and network-based QoE optimization	Caching optimization and CDN orchestration [28], server-assisted QoE delivery [1], bitrate guidance [29–31], bandwidth reservation [32–34] or QoE-driven video stream prioritization [35]	Calculate an optimal video bitrate to obtain QoE-fairness for each client. Some of these approaches assign different QoE-priority or QoE-policy to the requested video stream
Personalized QoE-centric Control, management, and QoE-fairness and QoE-fairness	[29, 36–39, 40–43]	Maximize the end-user's QoE by allocating resources efficiently to avoid unfair bandwidth sharing/slicing, video quality freezes, and instability
QoE-centric re-routing strategies	[44–49]	Compute the best delivery path that maximizes the user's QoE for multimedia services (e.g. using network links with and high system throughput and low packet loss and delay)
QoE-aware cross-layer optimization	[50–53]	Optimize the end-user's QoE by using a joint cooperation between layers and coordination of their actions during resources allocations. For example, the SDN controller can manage the QoE requirements from users that are specified at the application layer [54, 55]
Multimedia transport level optimizations	MPQUIC/MPTCP solutions for DASH [17, 24, 56, 57]	Improve the video QoE of DASH streams by transmitting subflows over multiple paths in the network. MPTCP and Multipath QUIC [58] provides quicker start of media streams, better network resources utilization, and higher network throughput that leads to better video streaming experience to users [19, 21]

and/or QUIC to improve the omnidirectional video streaming in lossy environments over SDN. The clients' initial delay is reduced up to 60% using an SDN application that is linked with Google Virtual Reality (VR) Viewer designed to reduce the client initialization delay by up to 60% in lossy networks. Hussein et al. [27] propose a QUIC-aware architecture that can handle load balancing, security at higher bandwidth utilization, and control network congestion in SDN-based networks. However, apart from the initial connection, the proposed architecture does not provide a justification of improving the video delivery for DASH systems. Table 5.1 provides a summary of QoE-Centric management strategies of multimedia streaming services in future networks.

5.3 Server and Network-Assisted Optimization Approaches Using SDN/NFV

DASH is the video streaming technology today used by major OTT service providers such as Hulu, Netflix, Amazon Video, and YouTube/YouTube. However, the performance of DASH becomes poor when multiple clients sharing the same bottleneck network link are streaming a video content from the server. Some of the streaming performance problems of DASH include video freezes, QoE-unfairness, high video quality switches, low bitrate, and underutilization of the network resources [39, 59]. These issues can be eliminated by the Server and Network-assisted DASH (SAND) approach that enhances the streaming performance by relying on standardized messages between DASH clients and network elements. SAND provides the performance and status of DASH clients as well as real-time information from the video servers, networks, cache, and CDNs [2]. The DASH Assisting Network Elements (DANEs) which are aware of network traffic can improve the streaming performance DASH by optimizing the bottleneck network links. With SAND, it is possible to partition or slice the network resources between DASH players and improve the end-user's QoE. The use case shown in Figure 5.3 employs SAND technology to enable service providers and operators for enhancing the end-user's video streaming experience while maintaining an efficient utilization of network bandwidth.

The DASH client running the DASH.js[3] can be embedded with QoE prediction, buffer filling strategy, and video bitrate and guidance algorithms. These algorithms enable a DASH client to learn about the availability of media content and types, video resolutions, and/or minimum and maximum bandwidths. Using this information, the DASH client selects the appropriate encoded video and starts streaming the content by fetching the segments using HTTP1/2/3 GET requests.

3 https://github.com/Dash-Industry-Forum/dash.js/wiki.

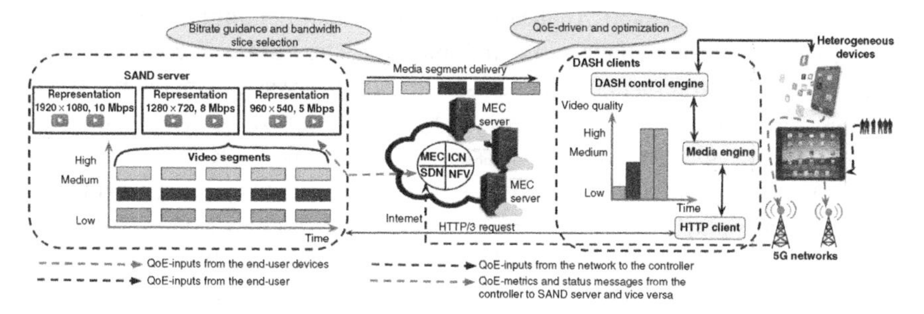

Figure 5.3 QoE-driven management of multimedia services over softwarized 5G networks.

Based on the predicted network throughput at each streaming point, the DASH client continues fetching the subsequent segments while monitoring in real time the network bandwidth fluctuations using the SDN controller. The DASH client fetches video segments as guided by bandwidth and QoE prediction schemes [60] by defining an appropriate bitrate levels in SDN-enabled adaptive streaming in future networks. Using SAND, the SDN controller can send information to the DASH clients about the video segment and the network throughput availability, as well as caching status of the segments. The arrows show QoE input information from the network to the controller and communication message exchanges to the video server.

Following the principles of SAND, many attempts have been made to improve the end-user's QoE using the bandwidth reservation and bitrate guidance mechanisms [29, 32, 33, 38, 42]. Mansy et al. [34] propose a network-level QoS strategy that can achieve QoE-fairness among competing clients sharing the limited network resources. Authors compute a video bitrate allocation algorithm that can achieve QoE maximal fairness between competing clients for the home network environments. To maximize the QoE for each client, using the proposed algorithm, the network controller considers the client's device screen size and video resolution to select the best video bitrate for downloading. The developed network-layer QoE-fair system consists of two important modules, namely the session manager and the bandwidth manager. The session manager calculates the QoE-fair bandwidth for clients and records active video streaming sessions. The bandwidth manager is responsible for allocating the minimum bandwidth to each video stream and enforcing QoE-fair allocations as calculated by the session manager.

A QoE-aware Bandwidth Management Solution (BMS) that can improve the end-user's QoE with different device capabilities (e.g. screen resolution, memory, and CPU) over SDN-based Hybrid Fiber Coax (HFC) access networks is proposed by Bentaleb et al. [33]. The three-layered BMS architecture consists of the control, application, and infrastructure layers. The application layer implements a Bandwidth Management Application (BMA) which communicates with the SDN controller through the northbound interface. The BMA also communicates with a set of video servers, HAS clients, and Multiple System Operators (MSO) entities. The BMA monitors the network by collecting both network and client statuses during video streaming and assists clients to make ABR decisions. It also detects system events such as playback state changes, clients joining/leaving the system, or their playback state. The heart of the BMS is the Viewer QoE Optimizer (VQO) that optimizes the end-user's QoE by computing their optimal joint video presentation decisions. The VQO specifically performs the following tasks: It determines the optimization of video streaming granularity for a group or single session (e.g. unicast VoD or OTTP services). The VQO formulates the bandwidth allocation by using the concave Network Utility Maximization (NUM) function to maximize the viewer's QoE. Moreover, the VQO employs the SDN Optimization Language (SOL) [61], a collection of models and methods such as the fast Model Predictive Control (fastMPC) [62], and the optimal online decomposition [63] to solve the overall QoE optimization problem of the end users.

Liotou et al. [30] propose a video bitrate-guided QoE-aware SDN application namely (*QoE-SDN APP*) for HTTP adaptive video streaming. The QoE-SDN-APP relies on SDN paradigm to perform QoE-aware video segment selection and caching mechanisms using the network exposure feedback from Video Service Providers (VSPs) and MNOs. The main component of the QoE-SDN APP is the QoE assessment module that determines and recommends the caching and encoding rate strategy that has to be used by VSP by considering the user mobility and future network load [2]. This module also uses MOS scale and Key performance Indicators (KPIs) (e.g. buffering events) to determine the QoE per application and allocates network resources so as to maximize the perceived QoE for the users. The SDN controller implements the VSP QoE control agent which allows the collaboration between the VSPs and the underlying MNO's infrastructure. It is worth mentioning that the QoE-SDN-APP enables VSPs to enhance quickly their video segment encoding and distribution procedures based on the network feedback exposed by the MNOs.

Kleinrouweler et al. [64] propose an SDN DASH-aware networking architecture that provides two mechanisms for explicit video adaptation assistance. The proposed system can guide DASH players and allocate network resources dynamically to provide an enhanced QoE. That way, the DASH players can reach their optimal video bitrate levels using the defined QoS/QoE policy and bitrate guidance mechanisms. The authors show that the number of video quality switches is greatly reduced and the optimal video bitrate can be doubled. That way, the end-user's QoE is significantly improved and allows network administrators to define shared QoE-based rules between video players. Bentaleb et al. [38] extended the idea in [64] by introducing SDNDASH, a novel SDN-based dynamic resource allocation and management architecture for DASH systems that overcomes quality instability, network resource underutilization and unfair bandwidth sharing between clients. A Resource Management Application (RMA) is developed to perform QoE-optimization, resource allocation, and monitoring in SDN-based platform using the RYU controller. That way, Software Defined Networking HTTP Adaptive Streaming (abbSDNHAS) can guide each client for the optimal video quality based on the bitrate level of the next chunk to be downloaded during video streaming. The average video quality, startup delay, number of stalls, and number of video switches are the parameters used by SDNHAS QoE model. Moreover, the authors use SSIMPlus objective quality measurement to represent the quality value of downloaded chunks. Compared to other approaches such as dash.js, SDNHAS can achieve 22% fewer bitrate switches and improves the QoE of DASH clients by 28%.

The authors in [32] compare the performance of bitrate guidance and bandwidth reservation techniques. The optimal bitrates for DASH clients are computed by the SDN controller to achieve video quality fairness, which is calculated based on the Structural SIMilarity (SSIM) index. When *bandwidth reservation* is used, the bandwidth slice is assigned to clients with similar optimal bitrates. As shown in Figure 5.4a, two control loops, *inner and outer control* loops, are used. Based on the video client feedback and bandwidth estimates, the inner control loop running at the client side selects the video bitrate. The outer control loop is executed in the network and sets the bandwidth slice. In the bitrate guidance scenario shown in Figure 5.4b, the optimal bitrates are computed by a centralized algorithm running in a network element. The video bitrates are then communicated to the DASH client that downloads the corresponding video segment. Guillen et al. [65] have introduced recently SAND/3, an SDN-based QoE control strategy for DASH over HTTP/3 that can orchestrate network-transport, and user-level components to ensure that the end-user's QoE is achieved. SAND/3 combines user, device, service, and network-level information. The SAND/3 architecture consists of

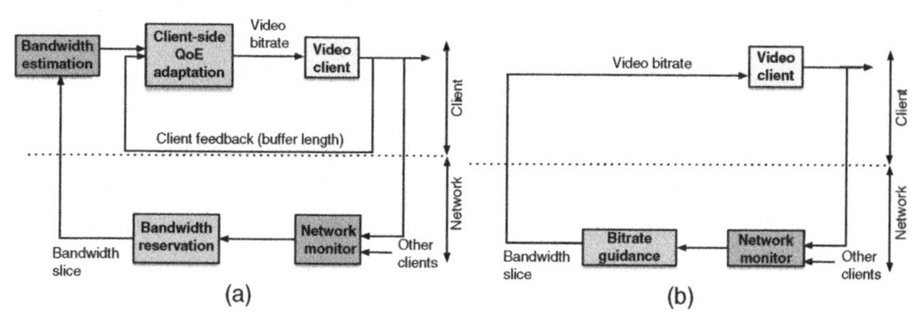

Figure 5.4 QoE-based network-assisted approaches for adaptive video streaming. (a) A specific bandwidth for each client is enforced by the network when bandwidth reservation is used, and (b) the network provides an explicit bitrate to the clients using bitrate guidance. Source: Cofano et al. [32].

three modules, namely, user, network, and QoE manager modules. The user module is responsible for collection user's identity. It also collects the devices' characteristics from a third-party service where the user is subscribed such as from Netflix or YouTube. The network module monitors the system events continuously and collects statistics from all entities in the network. Using MPTCP, it then uses the most suitable paths that offer the best QoS to route traffic. That way, it allows a better estimation of the available network resources to allocate them fairly to DASH clients. The QoE manager module uses the user's profile information, the current state of the network, and the specific service policies to recommend the suitable QoE level to the end users. Compared to TCP-based or QUIC-based only approaches, SAND/3 reduces significantly the number of stalls and improves the overall QoE of DASH clients.

5.4 QoE-Centric Fairness and Personalized QoE Control

This section presents QoE-fairness and personalized QoE-centric control mechanisms in SDN and/or NFV. Most of the approaches presented in this part employ the SDN controller to control and manage the video streaming performance of DASH players. The controller can monitor the streaming statuses, subscription plan types, and device capabilities of DASH players. Moreover, the controller has knowledge of the requested video content and buffer level as well as clients joining/leaving the network and starting/stopping the video playback. Previous approaches in this category such as [29, 38] optimize the QoE-fairness of competing clients using an optimization function that allocates the required video bitrate for DASH clients. It is important to mention that some of the previous approaches (e.g. [32]) discussed in Section 5.3 also address QoE-fairness using SAND architecture. This is different from the approaches discussed in this section where the SDN controller is used to control and manage QoE requirements of the competing DASH clients.

Georgopoulos et al. [29] propose an OpenFlow-Assisted QoE Fairness Framework (QFF) that achieves a QoE fairness level for different numbers of users sharing the limited network resources in heterogeneous environment. QFF considers the requirements of users, device characteristics and the current status of the network to optimize the QoE for all video streaming devices. The network management and active resource allocation mechanisms are implemented using an OpenFlow which allows a vendor-agnostic functionality in multimedia networks. The Network Inspector and the Media Presentation Description (MPD) Parser modules are used to monitor the status of the network and DASH video streaming sessions. The QFF, in turn, allocates dynamically the network resources to each DASH client device to achieve a maximized users' QoE -fairness level in multimedia networks. The Optimization Function (OF) and a Utility Function (UF) are used by the QFF to provide the network intelligence. QFF's utility functions provide a model that is used for mapping the video bitrate at a specific resolution and the QoE perceived by the client. The OF uses the models that the UF offer to find the optimum set of video bitrates that ensures QoE fairness across all DASH clients in the network [29].

As an extension to [38], Bentaleb et al. [39] propose SDNHAS, an intelligent cluster-based streaming architecture over SDN that enables HAS players to make better adaptation decisions in order to optimize the end-user's QoE. SDNHAS can allocate efficiently network resources to competing DASH clients in changing conditions. It can also reduce communication overhead between the SDN core and HAS players during video streaming. In order to accommodate a large number of HAS players with low complexity, an algorithm that uses the Structural Similarity Index Plus metric (SSIMplus) [66] for perceptual quality is developed. Figure 5.5 indicates an optimizer component, the main entity in SDNHAS which constructs a specific data structure for each cluster, called the *per-cluster QoE policy* for each video segment that is downloaded. A standard per-cluster QoE policy aggregates together a set of players, whose QoE policies belong to the same cluster [39]. That way, using specific QoE-policy, each player can receive a fair allocation of bandwidth with a maximized QoE level. It is worth mentioning that SDNHAS provides intelligent network management and QoE-aware adaptive streaming delivery mechanisms that achieve a maximized QoE-fairness level among heterogeneous HAS players.

The authors in [37] propose a client-Driven Video Delivery (cDVD) based on the principles of SDN and Participatory Networking (PANE) [67] to provide stable video quality by maintaining low re-buffering ratios in an encrypted SDN-assisted environment. cDVD employs a client-driven network-level QoE

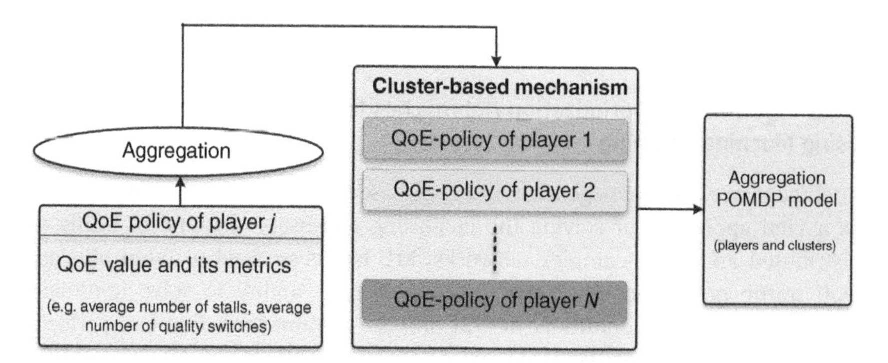

Figure 5.5 Per-cluster QoE policy structure and model abstraction. Source: Adapted from [39].

fairness strategy to enable the interaction between different DASH clients and the network components such as SDN controller. The controller uses the bandwidth enforcer to limit the bitrate of each video flow which in turn allocates bandwidth to maximize the end-user's QoE for all video streaming sessions. It is argued that it is possible to improve bottleneck utilization and client stability using the feedback and network-level QoE measurement for adaptive video streaming. It is important to mention that cDVD controls the bandwidth down-stream, which may be sufficient in many homes network. However, in general cases cDVD only provides upper bounds on the bandwidth to a video streaming.

Ramakrishnan et al. [40] propose a novel multi-client bandwidth management mechanism that optimizes the overall QoE of multiple video streaming sessions. The authors formulate the bandwidth allocation and joint video rate selection problem within a convex optimization SDN-based framework. The Video QoE optimization application (VQOA) module is implemented for collecting network level information at different points in the network. The VQOA utilizes this information to estimate accurately the end-user's QoE. The results indicate that with quality-optimized allocation the network can support more users with a QoE-fairness level of up to 75%. A user-level fairness model, UFair, which orchestrates network resource allocation between HAS streams to mitigate QoE fluctuations and improve the overall QoE fairness is given in [36]. UFair uses the video quality, switching impact, and cost efficiency metrics to measure the users' QoE fairness. Jiang et al. [41] propose an SDN-assisted multiplayer video streaming architecture that monitors the global network performance in real time. An online RL-based scheme is proposed to achieve QoE-user fairness level among multiple competing clients. The algorithm is implemented to guide the video bitrate selection and perform fair-bandwidth allocation to DASH clients. It is worth noting that, with the upcoming 5G networks, QoE-fairness [68] metrics should be used as the benchmark for QoE control and management of future multimedia services.

5.4.1 QoE-driven Optimization Approaches for Video Streaming Using Machine Learning

Intelligent network management and QoE-based monitoring solutions seem to be a vital approach for solving the increasing end-users' demand of traffic in distributed and large complex networks. ML has been used to build efficient QoE traffic prediction or control because of their ability to solve nonlinear problems [69–72]. The authors in [73] propose a framework that can perform video representation at the DASH client through a RL technique. The DASH controller is designed to perform an online learning of the temporal evolution of the system along with the algorithm that achieves a maximum video quality for the end users. The proposed algorithm based on the Markov Decision Process (MDP) optimization selects the best video representation while limiting both the buffering events and video quality variations in the system. Zhou et al. [74] propose mDASH, a Markov decision-based rate adaptation strategy for dynamic HTTP streaming services. In order to maximize the end-users' QoE, under changing conditions of the wireless channel, a greedy algorithm is proposed that considers important parameters including video rate switching frequency, buffer overflow/underflow, video playback quality, video amplitude, and buffer occupancy. Mao et al. [75] propose Pensieve, a system that can learn ABR algorithms using RL techniques that provide reward signals to reflect the video QoE based on the past streaming adaptation decisions. During video streaming, future video segments are selected following observations that are collected by the video players on the client side using the trained neural network model. Gadaleta et al. [76] propose a QoE optimization framework, namely D-DASH that can improve the end-users' QoE by combining the deep learning and reinforcement learning techniques. The video quality variations and freezing/rebuffering events parameters are taken into account by D-DASH during QoE optimization. Figure 5.6 shows a QoE-optimization and management using ML in softwarized networks. The network agent carries instructions from the SDN controller and translates them into a set of rules or policies on the data forwarding devices. An intelligent ML platform can learn from past experience, for example, from the video streaming data that is available based on the current state in the network. It uses predefined knowledge to forecast the load of incoming DASH requests which

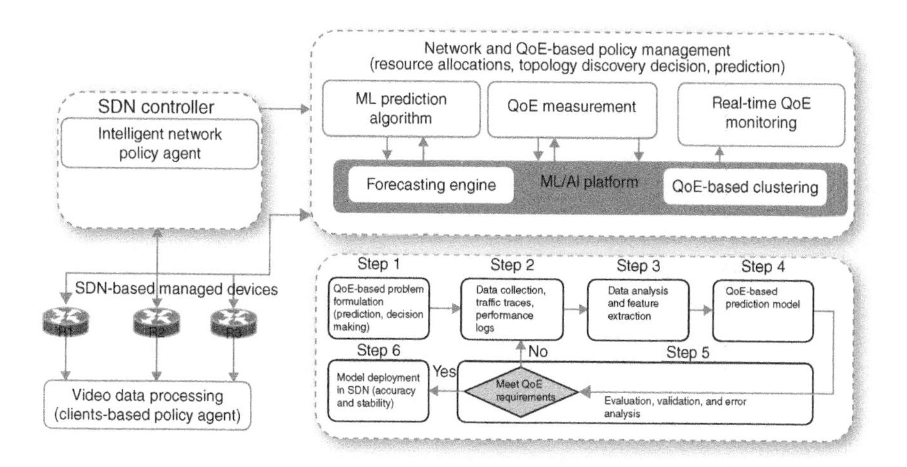

Figure 5.6 QoE-optimization and management using machine learning in softwarized networks.

in turn improves the overall performance of the multimedia streaming system. ML has been recently used for QoE prediction from network measurements. The QoE monitoring is responsible for monitoring the video quality in real time during a video streaming session. It monitors periodically different network parameters and communicates with QoE predictions module to forecast future video quality based on clients' device characteristics, bitrate, and changing network conditions. The QoE measurement module acquires network topology information and user's device capabilities. It also collects in real-time KQIs from the network and application during video streaming. To summarize, steps (1–6) indicating the chronological order of interactions for QoE measurements using ML are shown in Figure 5.6 using a flow-chart style.

A Fuzzy Logic Controller (FLC) that can perform adaptation of DASH video bitrate using both the estimated throughput and the client's buffer size is proposed in [77]. FLC minimize the ON-OFF switching during video streaming using the buffer dynamics and the observed network throughput to change the policy of

Table 5.2 A summary of QoE optimization approaches in future networks using ML.

References	Model	QoS/QoE parameters	Limitations and challenges
[72, 83]	Random Neural Networks (RNN)	Quantization Parameter (QP), Spatial Resolution (R), Packet Loss Rate (PLR), Mean Burst Length (MBL)	The implementation complexity and cost associated with QoE subjective feedback. Additional QoS parameters for dataset validation are needed
[84]	Artificial neural networks, support vector regression machines, decision trees, and Gaussian naive Bayes classifiers	Average delay, packet loss, average jitter, burst ratio, average burst size	Computational complexity does not consider economic factors like cost of a service
[85]	Random Neural Networks (RNN)	Content type	Does not consider rate control methods based on content types
[86]	Neural Network (NN) search method	Content type, screen resolutions, bottleneck capacity, delay, and packet loss	The design of new features for QoE prediction which are specific to video profiling [87] that can be measured by QoS monitoring systems are needed

choosing the next video bitrate that is to be downloaded. The concept of RL at client side is introduced in [78] where an agent is able to learn different configuration parameters under various network conditions. In order to reduce the average buffer filling, the decision process considers the bandwidth characteristics. This is achieved through an agent that perceives the feedback every time an action is taken using a reward function. Sieber et al. [79] propose an HTTP Adaptive Streaming using Machine Learning-based Adaptation Logics (HASBRAIN), a novel methodology for the training of ML-based adaptation algorithms for HAS. HASBRAIN can achieve a minimum number of video quality switches especially in a challenging mobile environment where the system throughput fluctuates regularly. Huang et al. [80] propose a video Quality Awareness Rate Control (QARC) that can achieve low latency and high video quality using DRL techniques. Using QARC, future video bitrates are selected by considering the previous network status and historical video downloaded frames. Huang et al. [81] introduce Tiyuntsong, a self-play RL method with Generative Adversarial Network (GAN) for ABR video streaming services. The authors in [82] propose a continual learning strategy for video bitrate selection in media streaming services called Fugu which can learn from its users and produce accurate QoE predictions. Using Fugu, the trained neural network can perform QoE prediction using the previous history and internal TCP statistics to make decisions for downloading the next video chunk. Table 5.2 provides a summary of QoE optimization approaches in future networks using ML.

5.5 Conclusion

It is evident that both academia and industry have made significant efforts to investigate different mechanisms for improving the end-user's QoE. This chapter provides the QoE management of multimedia streaming services using SDN and/or NFV. QoE-centric routing mechanisms are first presented followed by QoE optimization using MPTCP/SR and QUIC approaches. The chapter further provides server and network-assisted strategies for QoE to improve the end-user's QoE along with QoE-centric fairness and personalized QoE control using SDN. Moreover, QoE-driven optimization for video streaming services using ML is extensively explored.

Bibliography

1 Thomas, E., van Deventer, M.O., Stockhammer, T. et al. (2017). Enhancing MPEG DASH performance via server and network assistance. *SMPTE Motion Imaging Journal* 126 (1): 22–27.

2 Barakabitze, A.A., Barman, N., Ahmad, A. et al. (2019). QoE management of multimedia services in future networks: a tutorial and survey. *IEEE Communication Surveys and Tutorials* 22 (1): 526–565.

3 Egilmez, H.E., Civanlar, S., and Tekalp, A.M. (2013). An optimization framework for QoS-enabled adaptive video streaming over openflow networks. *IEEE Transactions on Multimedia* 15 (3): 710–715.

4 Sayit, M., Cetinkaya, C., Karayer, E., and Hellge, C. (2014). SDN for segment based flow routing of DASH. *4th International Conference on Consumer Electronics Berlin (ICCE-Berlin)*, pp. 74–77.

5 Dobrijevic, O., Santl, M., and Matijasevic, M. (2015). Ant colony optimization for QoE-centric flow routing in software-defined networks. *Proceedings of the 11th International Conference on Network and Service Management, CNSM 2015*, pp. 274–278. https://doi.org/10.1109/CNSM.2015.7367371.

6 Bouten, N., Claeys, M., Van Poecke, B. et al. (2016). Dynamic server selection strategy for multi-server HTTP adaptive streaming services. *12th International Conference on Network and Service Management (CNSM)*, pp. 82–90.

7 Al-Jawad, A., Shah, P., Gemikonakli, O., and Trestian, R. (2018). LearnQoS: A learning approach for optimizing QoS over multimedia-based SDNs. *Proceedings IEEE International Symposium on Broadband Multimedia Systems and Broadcasting (BMSB)*, pp. 1–6.

8 Calvigioni, G., Aparicio-Pardo, R., Sassatelli, L. et al. (2018). Quality of experience-based routing of video traffic for overlay and ISP networks. *IEEE International Conference on Computer Communications (INFOCOM)*, pp. 121–132.

9 Huang, X., Yuan, T., Qiao, G., and Ren, Y. (2018). Deep reinforcement learning for multimedia traffic control in software defined networking. *Computer Networks* 32 (6): 35–41.

10 Khalid, A., Zahran, A.H., and Sreenan, C.J. (2019). An SDN-based device-aware live video service for inter-domain adaptive bitrate streaming. *Proceedings of the 10th ACM Multimedia Systems Conference*, pp. 121–132.

11 Sayit, M., Cetinkay, C., Yildiz, H.U., and Tavli, B. (2019). DASH-QoS: A scalable network layer service differentiation architecture for DASH over SDN. *Computer Networks* 154 (8): 12–25.

12 da Costa Filho, R.I.T., Lautenschläger, W., Kagami, N. et al. (2018). Scalable QoE-aware path selection in SDN-based mobile networks. *2018 IEEE International Conference on Computer Communications (INFOCOM)*, pp. 1–6, April 2018.

13 Frnda, J., Voznak, M., and Sevcik, L. (2016). Impact of packet loss and delay variation on the quality of real-time video streaming. *Telecommunication Systems* 62 (2): 265–275. https://doi.org/10.1007/s11235-015-0037-2.

14 Venkataraman, M. and Chatterjee, M. (2014). Effects of internet path selection on video-QoE: analysis and improvement. *IEEE/ACM Transactions on Networking* 22 (3): 689–702.

15 Quang, P.T.A., Piamrat, K., Singh, K.D., and Cesar, V. (2017). Video streaming over ad hoc networks: a QoE-based optimal routing solution. *IEEE Transactions on Vehicular Technology* 66 (2): 1533–1546.

16 Sanchez-Iborra, R., Cano, M.-D., and Garcia-Haro, J. (2014). Performance evaluation of BATMAN routing protocol for VoIP services: a QoE perspective. *IEEE Transactions on Wireless Communications* 13 (9): 4947–4958.

17 Barakabitze, A.A., Mkwawa, I.-H., Sun, L., and Ifeachor, E. (2018). QualitySDN: Improving video quality using MPTCP and segment routing in SDN/NFV. *IEEE Conference on Network Softwarization*, May 2018.

18 Barakabitze, A.A., Sun, L., Mkwawa, I.-H., and Ifeachor, E. (2018). A novel QoE-centric SDN-based multipath routing approach of multimedia services over 5G networks. *IEEE International Conference on Communications*, May 2018 (Accepted).

19 Wu, J., Yuen, C., Cheng, B. et al. (2016). Streaming high-quality mobile video with multipath TCP in heterogeneous wireless networks. *IEEE Transactions on Mobile Computing* 15 (9): 2345–2361.

20 Mizuochi, T. (2010). Forward error correction. In: *High Spectral Density Optical Communication Technologies, Optical and Fiber Communications Reports, 6* (ed. M. Nakazawa, K. Kikuchi, and T. Miyazaki): Heidelberg: Springer-Verlag Berlin, 1–338.

21 Corbillon, X., Aparicio-Pardo, R., Kuhn, N. et al. (2016). Cross-layer scheduler for video streaming over MPTCP. *Proceedings of the 7th International Conference on Multimedia Systems*, May 2016.

22 James, C., Halepovic, E., Wang, M. et al. (2016). Is multipath TCP (MPTCP) beneficial for video streaming over DASH? *IEEE 24th International Symposium on Modeling, Analysis and Simulation of Computer and Telecommunication Systems (MASCOTS)*, pp. 331–336.

23 Han, B., Qian, F., Ji, L., and Gopalakrishnan, V. (2016). MP-DASH: Adaptive video streaming over preference-aware multipath. *Proceedings of the 12th International on Conference on emerging Networking EXperiments and Technologies - CoNEXT '16*, pp. 129–143.

24 Nam, H., Doru, C., and Henning, S. (2016). Towards dynamic MPTCP path control using SDN. *IEEE NetSoft Conference and Workshops (NetSoft)*, pp. 286–294, July 2016.

25 Mohan, N., Shreedhar, T., Zavodovski, A. et al. (2019). Is two greater than one: analyzing multipath TCP over dual-LTE in the wild. *arXiv preprint arXiv:1909.02601*.

26 Hayes, B., Chang, Y., and Riley, G. (2017). Omnidirectional adaptive bitrate media delivery using MPTCP/QUIC over an SDN architecture. *IEEE Global Communications Conference*, December 2017.

27 Hussein, A., Kayssi, A., Elhajj, I.H., and Chehab, A. (2018). SDN for QUIC: An enhanced architecture with improved connection establishment. *SAC '18: Proceedings of the 33rd Annual ACM Symposium on Applied Computing*, pp. 2136–2139, April 2018.

28 Ganjam, A., Jiang, J., Liu, X. et al. (2015). C3: Internet-scale control plane for video quality optimization. *Proceedings of the 12th USENIX Conference on Networked Systems Design and Implementation*, NSDI'15, pp. 131–144, Berkeley, CA, USA: USENIX Association. ISBN 978-1-931971-218.

29 Georgopoulos, P., Elkhatib, Y., Broadbent, M. et al. (2013). Towards network-wide QoE fairness using openflow assisted adaptive video streaming. *Proceedings of the 2013 ACM SIGCOMM Workshop on Future Human-centric Multimedia Networking*, pp. 15–20, February 2013.

30 Liotou, E., Samdanis, K., Pateromichelakis, E. et al. (2018). QoE-SDN APP: A rate-guided QoE-aware SDN-APP for HTTP adaptive video streaming. *IEEE Journal on Selected Areas in Communications* 36 (3): 598–615 (99): 1.

31 Bhat, D., Rizk, A., Zink, M., and Steinmetz, R. (2017). Network assisted content distribution for adaptive bitrate video streaming. *Proceedings of the 8th ACM on Multimedia Systems Conference*, pp. 62–75, June 2017.

32 Cofano, G., De Cicco, L., Zinner, T. et al. (2017). Design and experimental evaluation of network-assisted strategies for HTTP adaptive streaming. *ACM Transactions on Multimedia Computing, Communications, and Applications* 35 (11): 1–24.

33 Bentaleb, A., Begen, A.C., and Zimmermann, R. (2018). QoE-aware bandwidth broker for HTTP adaptive streaming flows in an SDN-enabled HFC network. *IEEE Transactions on Broadcasting* PP (99): 1–15.

34 Mansy, A., Fayed, M., and Ammar, M. (2015). Network-layer fairness for adaptive video streams. *IFIP Networking Conference (IFIP Networking)*, pp. 1–9, May 2015.

35 Petrangeli, S., Wu, T., Wauters, T. et al. (2017). A machine learning-based framework for preventing video freezes in HTTP adaptive streaming. *Journal of Network and Computer Applications* 94: 78–92.

36 Mu, M., Broadbent, M., Farshad, A. et al. (2016). A scalable user fairness model for adaptive video streaming over SDN-assisted future networks. *IEEE Journal on Selected Areas in Communications* 34 (8): 2168–2184.

37 Chen, J., Ammar, M., Fayed, M., and Fonseca, R. (2016). Client-driven network level QoE fairness for encrypted 'DASH-S'. *Proceedings of the 2016 Workshop on QoE-Based Analysis and Management of Data Communication Networks*, pp. 55–60, August 2016.

38 Bentaleb, A., Begen, A.C., and Zimmermann, R. (2016). SDNDASH: Improving QoE of HTTP adaptive streaming using software defined networking. *Proceedings of the ACM Conference on Multimedia*, pp. 1296–1305, October 2016.

39 Bentaleb, A., Begen, A.C., Zimmermann, R., and Harous, S. (2017). SDNHAS: An SDN-enabled architecture to optimize QoE in HTTP adaptive streaming. *IEEE Transactions on Multimedia* 19 (10): 2136–2151.

40 Ramakrishnan, S., Zhu, X., Chan, F. et al. (2016). Optimizing quality-of-experience for HTTP-based adaptive video streaming: an SDN-based approach. *International Journal of Multimedia Data Engineering and Management* 7 (4): 22–44.

41 Jiang, J., Hu, L., Hao, P. et al. (2017). Q-FDBA: Improving QoE fairness for video streaming. *Multimedia Tools and Applications* 77: 10787–10806.

42 Kleinrouweler, J.W., Cabrero, S., and Cesar, P. (2016). Delivering stable high-quality video: an SDN architecture with DASH assisting network elements. *7th International ACM Conference on Multimedia Systems*, May 2016.

43 Wang, Y., Li, P., Jiao, L. et al. (2016). A data-driven architecture for personalized QoE management in 5G wireless networks. *IEEE Wireless Communications Magazine* 24 (1): 102–110.

44 Dobrijevic, O., Kassler, A.J., Skorin-Kapov, L., and Matijasevic, M. (2014). Q-POINT: QoE-driven path optimization model for multimedia services. *Wired/Wireless Internet Communications* 8458: 134–147.

45 Dobrijevic, O., Santl, M., and Matijasevic, M. (2015). Ant colony optimization for QoE-centric flow routing in software-defined networks. *Proceedings of the 11th International Conference on Network and Service Management (CNSM)*, pp. 274–278, November 2015.

46 Yang, J., Yang, E., Ran, Y., and Chen, S. (2015). *SDM^{2}Cast*: An openflow-based, software-defined scalable multimedia multicast streaming framework. *IEEE Internet Computing* 19 (4): 36–44.

47 Athanasopoulos, D., Politis, I., Lykourgiotis, A. et al. (2016). End-to-end quality aware optimization for multimedia clouds. *2016 International Conference on Telecommunications and Multimedia (TEMU)*, pp. 1–5, July 2016. https://doi.org/10.1109/TEMU.2016.7551931.

48 Mkwawa, I.-H., Barakabitze, A.A., and Sun, L. (2016). Video quality management over the software defined networking. *IEEE International Symposium on Multimedia (ISM)*, pp. 559–564, December 2016.

49 Grigoriou, E., Barakabitze, A.A., Atzori, L. et al. (2017). An SDN-approach for QoE management of multimedia services using resource allocation. *IEEE International Conference Communications (ICC)*, May 2017.

50 Zhao, M., Gong, X., Liang, J. et al. (2015). QoE-driven cross-layer optimization for wireless dynamic adaptive streaming of scalable videos over HTTP. *IEEE Transactions on Circuits and Systems for Video Technology* 25 (3): 451–465.

51 Yang, M., Li, Y., Hu, L. et al. (2015). Cross-layer software-defined 5G network. *Mobile Networks and Applications* 20: 400–409.

52 Huang, W., Ding, L., Meng, D. et al. (2018). QoE-based resource allocation for heterogeneous multi-radio communication in software-defined vehicle networks. *IEEE Access* 6: 3387–3399. https://doi.org/10.1109/ACCESS.2018.2800036.

53 De Souza, F.R., Miers, C.C., Fiorese, A., and Guilherme, G.P. (2017). QoS-aware virtual infrastructures allocation on SDN-based clouds. *17th IEEE/ACM International Symposium on Cluster, Cloud and Grid Computing*, pp. 120–129, Madrid, Spain, May 2017.

54 Awobuluyi, O., Nightingale, J., Wang, Q., and Alcaraz-Calero, J.M. (2015). Video quality in 5G networks context-aware QoE management in the SDN control plane. *2015 IEEE International Conference on Computer and Information Technology; Ubiquitous Computing and Communications; Dependable, Autonomic and Secure Computing; Pervasive Intelligence and Computing*, pp. 1657–1662, October 2015.

55 Liotou, E., Tsolkas, D., Passas, N., and Merakos, L. (2015). Quality of experience management in mobile cellular networks: key issues and design challenges. *IEEE Communications Magazine* 53 (7): 145–153.

56 Barakabitze, A.A., Sun, L., Mkwawa, I.-H., and Ifeachor, E. (2018). A novel QoE-centric SDN-based multipath routing approach of mutimedia services over 5G networks. *IEEE International Conference on Communications*, May 2018.

57 Herguner, K., Kalan, R.S., Cetinkaya, C., and Sayit, M. (2017). Towards QoS-aware routing for DASH utilizing MPTCP over SDN. *IEEE Conference on Network Function Virtualization and Software Defined Networks (NFV-SDN)*, pp. 1–6, December 2017.

58 Arisu, S., Yildiz, E., and Begen, A.C. (2019). Game of protocols: is QUIC ready for prime time streaming? *International Journal of Network Management* e2063: 1–18.

59 Rodríguez, D.Z., Wang, Z., Rosa, R.L., and Bressan, G. (2014). The impact of video-quality-level switching on user quality of experience in dynamic adaptive streaming over HTTP. *EURASIP Journal on Wireless Communications and Networking* 2014 (1): 1–15.

60 Al-Issa, A., Bentaleb, A., Barakabitze, A.A. et al. (2019). Bandwidth prediction schemes for defining bitrate levels in SDN-enabled adaptive streaming. *International Conference on Network and Service Management (CNSM)*, October 2019.

61 Heorhiadi, V., Reiter, M.K., and Sekar, V. (2016). Simplifying software-defined network optimization using SOL. *Proceedings of the 13th USENIX Conference on Networked Systems Design and Implementation*, NSDI'16, pp. 223–237, Berkeley, CA, USA: USENIX Association. ISBN 978-1-931971-29-4. http://dl.acm.org/citation.cfm?id=2930611.2930627 (accessed 3 May 2022).

62 Wang, Y. and Boyd, S. (2010). Fast model predictive control using online optimization. *IEEE Transactions on Control Systems Technology* 18 (2): 267–278. https://doi.org/10.1109/TCST.2009.2017934.

63 Palomar, D.P. and Chiang, M. (2006). A tutorial on decomposition methods for network utility maximization. *IEEE Journal on Selected Areas in Communications* 24 (8): 1439–1451. https://doi.org/10.1109/JSAC.2006.879350.

64 Kleinrouweler, J.W., Cabrero, S., and Cesar, P. (2016). Delivering stable high-quality video: an SDN architecture with DASH assisting network elements. *Proceedings of the 7th International Conference on Multimedia Systems (MMSys)*, May 2016.

65 Guillen, L., Izumi, S., Abe, T., and Suganuma, T. (2019). SAND/3: SDN-assisted novel QoE control method for dynamic adaptive streaming over HTTP/3. *Electronics* 8 (8): 864.

66 Rehma, A., Zeng, K., and Wang, Z. (2015). Display device-adapted video quality-of-experience assessment. *Proceedings of SPIE-The International Society for Optical Engineering*, 939406.

67 Ferguson, A.D., Guha, A., Liang, C. et al. (2013). Towards network-wide QoE fairness using openflow assisted adaptive video streaming. *Proceedings of the ACM SIGCOMM 2013*, pp. 327–338. August 2013.

68 Hoßfeld, T., Skorin-Kapov, L., Heegaard, P.E., and Varela, M. (2017). Definition of QoE fairness in shared systems. *IEEE Communications Letters* 21 (1): 184–187.

69 Kang, Y., Chen, H., and Xie, L. (2013). An artificial-neural-network-based QoE estimation model for video streaming over wireless networks. *IEEE/CIC International Conference on Communications in China (ICCC)*, Xi'an, China, pp. 264–269.

70 Wang, C., Jiang, X., Menga, F., and Wang, Y. (2011). Quality assessment for MPEG- 2 video streams using a neural network model, pp. 868–872.

71 Staelens, N., Van Wallendael, G., Crombecq, K. et al. (2012). No-reference bitstream-based visual quality impairment detection for high definition H.264/AVC encoded video sequences. *IEEE Transactions on Broadcasting* 58 (2): 187–199.

72 Danish, E., Alreshoodi, M., Fernando, A. et al. (2016). Cross-layer QoE prediction for mobile video based on random neural networks. *2016 IEEE International Conference on Consumer Electronics (ICCE)*, pp. 227–228, Las Vegas, USA.

73 Chiariotti, F., Toni, L., D'Aronco, S., and Frossard, P. (2016). Online learning adaptation strategy for DASH clients. *Proceedings of the 7th International Conference on Multimedia Systems*, Volume18, pp. 738–751.

74 Zhou, C., Lin, Z., and Guo, C.W. (2016). mDASH: A Markov decision-based rate adaptation approach for dynamic HTTP streaming. *IEEE Transactions on Multimedia* 18: 738–751.

75 Mao, H., Netravali, R., and Alizadeh, M. (2017). Neural adaptive video streaming with pensieve. *Proceedings of the Conference of the ACM Special Interest Group on Data Communication.*

76 Gadaleta, M., Chiariotti, F., Rossi, M., and Zanella, A. (2017). D-DASH: A deep Q-learning framework for DASH video streaming. *IEEE Transactions on Cognitive Communications and Networking* 3 (4): 703–718.

77 Sobhani, A., Yassine, A., and Shirmohammadi, S. (2015). A fuzzy-based rate adaptation controller for DASH. *Proceedings of the 25th ACM Workshop on Network and Operating Systems Support for Digital Audio and Video*, pp. 31–36, March 2015.

78 van der Hoofty, J., Petrangeliy, S., Claeysy, M. et al. (2015). A learning-based algorithm for improved bandwidth-awareness of adaptive streaming clients. *IFIP/IEEE International Symposium on Integrated Network Management (IM)*, May 2015.

79 Sieber, C., Hagn, K., Moldovan, C. et al. (2018). Towards machine learning-based optimal has, August 2018. *arXiv:1808.08065.*

80 Huang, T., Zhang, R.-X., Zhou, C., and Sun, L. (2018). QARC: Video quality aware rate control for real-time video streaming based on deep reinforcement learning. *Proceedings of the 26th ACM International Conference on Multimedia*, October 2018.

81 Huang, T., Yao, X., Wu, C. et al. (2019). Tiyuntsong: a self-play reinforcement learning approach for ABR video streaming, May 2019. *arXiv:1811.06166.*

82 Yan, F.Y., Ayers, H., Zhu, C., and Fouladi, S. (2019). Continual learning improves internet video streaming, September 2019. *arXiv:1906.01113.*

83 Danish, E., Fernando, A., Alreshoodi, M., and Woods, J. (2016). A hybrid prediction model for video quality by QoS/QoE mapping in wireless streaming. *IEEE International Conference on Communication Workshop (ICCW)*, London, UK, pp. 1723–1728.

84 Tsamardinos, I., Charonyktakis, P., Plakia, M., and Papadopouli, M. (2016). On user-centric modular QoE prediction for VoIP based on machine-learning algorithms. *IEEE Transactions on Mobile Computing* 15 (6): 1443–1456.

85 Ghalut, T., Larijani, H., and Shahrabi, A. (2015). Content-based video quality prediction using random neural networks for video streaming over LTE networks. *IEEE International Conference on Computer and Information Technology; Ubiquitous Computing and Communications; Dependable, Autonomic* *and Secure Computing; Pervasive Intelligence and Computing*, Liverpool, UK, pp. 1626–1631.

86 Paris, S., Maggi, L., Debbah, M. et al. (2019). Predicting QoE factors with machine learning. *IEEE International Conference on Communications (ICC)*, May 2019.

87 Tsilimantos, D., Karagkioules, T., Nogales-Gomez, A., and Valentin, S. (2017). Traffic profiling for mobile video streaming. *IEEE International Conference on Communications (ICC)*, May 2017.

6

Network Softwarization and Virtualization in Future Networks: The promise of SDN, NFV, MEC, and Fog/Cloud Computing

6.1 Network Softwarization: Concepts and Use Cases

Softwarization has been dubbed a crucial characteristic of 5G networking since it shifts network function support from traditional hardware-based solutions to software-based solutions [1]. The use of software programming for the design, implementation, deployment, and maintenance of network equipment/components/services is known as network softwarization [2]. The goal of network softwarization is to make 5G services and applications more agile and cost-effective [3]. Network softwarization promises to provide E2E service management and control while guaranteeing QoE of the end users. Network programmability, flexibility, and adaptability are some of the 5G requirements offered by network softwarization along with the slicing as-a-service and the overall 5G E2E service platform unification [4]. Important components and enablers of network softwarization include Software-Defined Networking (SDN), Network Function Virtualization (NFV), Multiaccess Edge Computing (MEC), Service Function Chaining (SFC), network slicing, and network virtualization, and the transition from today's network of entities toward a "network of functions." The network functions in the context of softwarization are composed on an "on-demand" and "on-the-fly" basis. Programmability in network softwarization allows for dynamic changes in the functionality of some of network components. Dynamic programmability to network equipment such as routers, switches application servers, and all segments of the network make it easier for network softwarization to introduce new network services. It also enables the rapid, flexible, and dynamic deployment of new network and management services as groups of Virtual Machines (VMs) in the data plane, control plane, management plane, and service plane [1, 4].

Multimedia Streaming in SDN/NFV and 5G Networks: Machine Learning for Managing Big Data Streaming, First Edition. Alcardo Barakabitze and Andrew Hines.

New design and implementation in various 5G network segments are required to accomplish network softwarization goals because each segment (such as mobile-edge networks, core networks, network clouds, radio access network (RAN), and transport networks) has different technical characteristics, requirements, and level of softwarization [5]. Figure 6.1 shows the main layers software-defined network architecture and its applications in the context of QoE control and management for multimedia streaming services. By extending the traditional data center to the edge of 5G networks, softwarization in mobile edge networks attempts to bring content, network services, and resources closer to the end user. Softwarization will be deployed in mobile edge networks using a virtualized architecture that combines SDN, Information-Centric Networking (ICN), NFV, and MEC [6, 7]. Network softwarization of MEC in 5G promises to reduce the amount of data transported to the 5G core network for processing while enabling real time and application flow information. It will also make efficient use of available network resources by caching video streaming contents at the MEC server, a concept similar to ICN [4]. Following the SDN/NFV architectural principles, network softwarization in core networks should be implemented as Virtual Network Functions (VNFs) that run on VMs enabled in Fog/Cloud Computing (CC) environments, potentially over standard servers [8]. Network softwarization in the transport network is designed to be implemented using appropriate interfaces in SDN/NFV environments where network optimization and resource discovery can be easily implemented in the 5G control plane [1, 9].

6.2 Network Softwarization and Virtualization Technologies for Future Communication Platforms

6.2.1 Software-Defined Networking (SDN)

The SDN enables softwarization in next-generation network by decoupling the data plane and control plane. The data plane represents the network devices. The network management policies, automation, control, and optimization is deployed in the application plane connected to the control plane by NorthBound Interface (NBI) [4]. SDN is defined by the Open Network Foundation (ONF) as "*the physical separation of the network control plane from the forwarding plane, and where a control plane governs many devices.*" With a global view of the entire network, this separation provides flexibility and centralized control. It also has the ability to react quickly to changing network conditions, business, market, and end user's requirements.

Moreover, network softwarization offered by SDN allows a global view of the network for network management and monitoring which enhances the optimization of the network resources. The SDN also allows the automation of the control

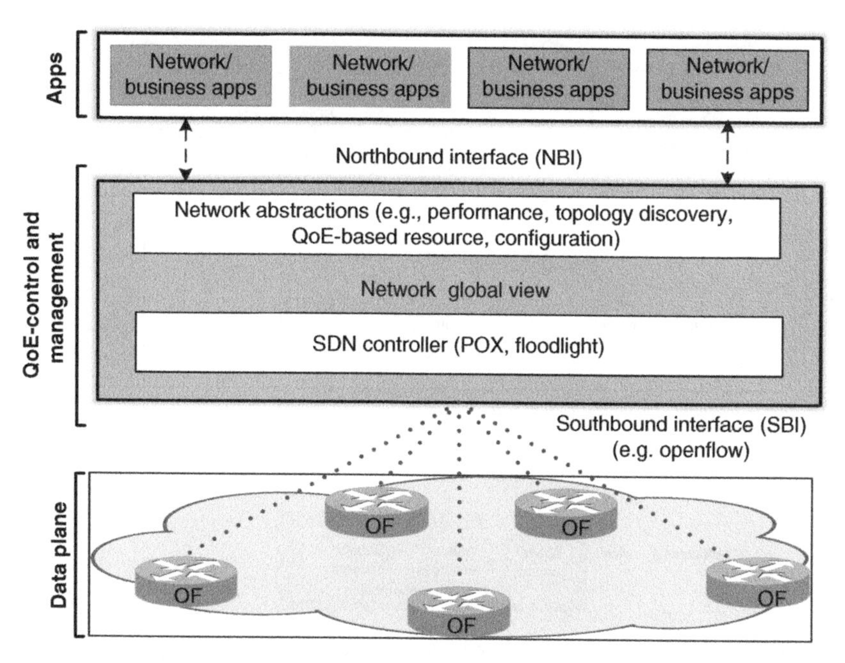

Figure 6.1 Main layers of the SDN architecture.

and network optimization together with the deployment of the data-driven approaches on top of the SDN controller in the application plane, which indeed makes future networks more adaptive, data-driven, with dynamic network configuration, management, and control. The network control in SDN architecture is directly programmable using SBI such as OpenFlow [10], FoRCES [11], and OpFlex [12]. It is worth noting that the data forwarding plane can run on a specialized commodity server [13] (e.g. VMware's NSX platform [14]) consisting of the open virtual switch (vSwitch) and the SDN controller. The design and implementation of an SDN controller can be either centralized (for managing small networks or a single administrative domain, for example Floodlight [15], Beacon [16], and Ryu NOS [17]) or distributed controllers, for example Hyperflow [18], HP VAN SDN [19], DISCO [20], and ONOS [21] (for overcoming the reliability and scalability issues in large-scale networks that span multiple control domains).

The ONF has developed a thorough SDN paradigm for 5G network slicing analysis [22] as shown in Figure 6.2, where the context of every SDN client context in the ONF architecture shows a potential slice. The SDN controller uses a collection of rules or policies to manage network slices. The server and client contexts as well as the installation of different policies are facilitated by the SDN controller [23]. The SDN controller, in particular, keeps track of network

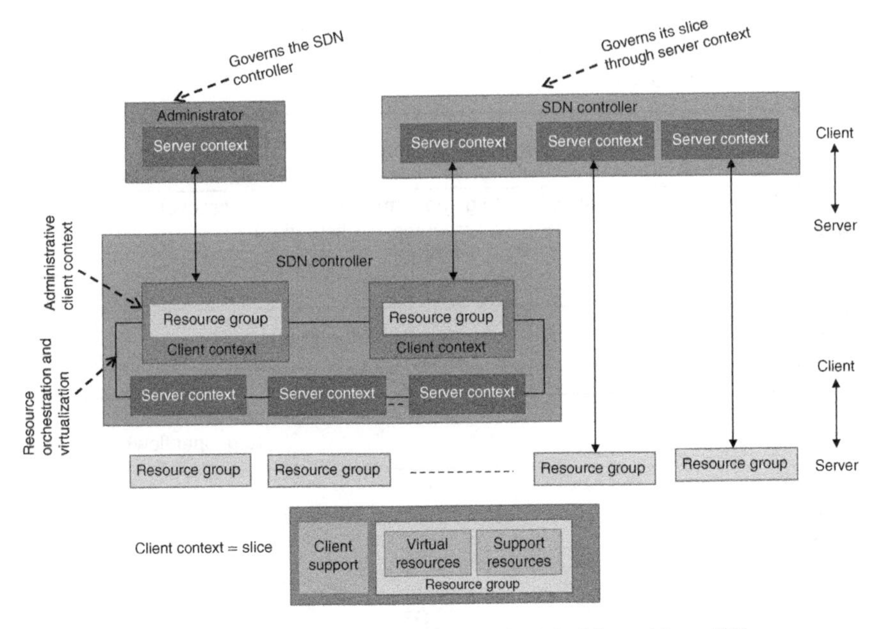

Figure 6.2 ONF SDN network slicing architecture. Source: Adapted from [22] .

slice client context. As a result, an SDN controller can manage network slices dynamically by grouping slices that belong to the same context [24]. On the server context, the SDN controller governs its slices and executes resource orchestration. To satisfy any incoming demands from end users, the client context comprises support, client, and virtual resources [4].

6.2.2 Standardization Activities of SDN

The 3GPP, ONF, IEEE, ETSI, and the Internet Engineering Task Force (IETF) are the major standardization bodies working on the standardization activities of SDN including the development of the SBI (OpenFlow standards) and definitions of interfaces and protocols for network infrastructure abstraction. Some of the bodies and consortia have been working on development of technical elements of SDN (e.g. scaling solutions, abstractions, security, performance, reliability requirements, frameworks/architectures, and models and programming languages). The main building blocks of SDN that are established by the open source software communities include CloudStack [25], OpenStack [26], and OpenDaylight [27] driving adoption and promoting innovation.

The standardization activities regarding SDN within the ITU span across different study groups (SGs). For example, the network protocols and signaling requirements of SDN for Wireless Access Network (WAN) services and bandwidth adjustment on the broadband network gateway have been investigated by the SG11. The SG13 specifies requirements, use cases, and architecture of SDN for future softwarized betworks. The open source software community such as CloudStack [25], OpenDaylight [27], and OpenStack [26] have been developing, fostering innovation, and accelerating the adoption the basic building blocks of SDN. The Metro Ethernet Forum (MEF) [28] introduced SDN technologies to the carrier Ethernet services. Continually evolving, the MEF facilitated the industry-neutral implementation environments for service orchestration and Layer 2 – Layer 7 connectivity services based on open source SDN and NFV.

6.2.3 Traffic Management Applications for Stateful SDN Data Plane

Despite the fact that the OpenFlow standard includes several flow tables in the OpenFlow pipeline, state information in the SDN data plane cannot be maintained. OpenFlow also heavily relies on the SDN controller to keep track of all packet statuses. The basic match/action abstraction of OpenFlow switches, on the other hand, limits the controller's ability to fully supervise the growth of the forwarding rules. This can be particularly restricting for a variety of applications, such as traffic management applications, that are influenced by the slow control path's delay [29]. Because of the control channel bottleneck and processing latency imposed between the SDN controller and switches, the static nature of the OpenFlow forwarding abstraction could present scalability, reliability, and security issues in 5G network slicing [30]. To offload states and operation logics from the controller to the data plane, a stateful data plane is proposed. Switches with a stateful data plane can conduct some tasks separately, which speeds up packet processing while lowering controller and network overhead. However, the stateful data plane adds to the complexity of network devices and introduces a slew of new administration and scheduling difficulties for SDN-enabled networks. Thanks to the advanced switch interface technologies such as P4 [31], OpenState [32], Stateful Data Plane Architecture (SDPA) [33], POF [34], and SNAP [35] that offer enhanced Stateful forwarding and expose persistent state on the data forwarding plane [32]. It is worth mentioning that the network slices of future softwarized networks must be regulated, monitored, and managed autonomously using stateful forwarding technologies, while supporting a variety of protocols and data transport mechanisms [36] (Table 6.1).

6.2.4 Network Function Virtualization (NFV)

The network virtualization enabled by NFV provides several opportunities in future networks in terms of deployment options, scalability, and agility of the

Table 6.1 A summary of SDN open source platforms, standardization efforts, projects, and implementations.

Name	Description	Working group	Description of SDN-related work
ONF	Industry-led Consortia for OpenFlow Standardization	–	Analyze SDN requirements, OpenFlow Standard regarding to SDN concepts, frameworks, architecture, software, and certifications.
OpenDaylight	Industry-led Consortia	ETSI NFV	To accelerate the adoption and fostering new innovation for SDN applications and services.
ITU-T	International Organization	SG (11,13, 15 & 17)	Investigating the functional architecture, signaling requirements, and protocols for SDN (e.g. security and unified intelligent programmable interface for IPv6).
IETF/IRTF	IRTF WG	SDN RG	To develop a customizable framework capable of orchestrating network services across heterogeneous NFV infrastructures.

network. The NFV allows the deployment of NFs such as routers, switches, firewalls, and load balancer on top of commodity hardware (computer servers) as VNFs instead of Physical Network Functions (PNFs) [38]. Thus, using NFV in future networks, any NF can be added, removed, or updated within the network on-demand using general computing infrastructure which leads to dynamic scaling of SFC (Table 6.2). The major key advantage of the NFV is the automation of the VNFs/components deployment using ETSI NFV Management and Orchestration (MANO) architecture [4]. This will allow data-driven network management application to be deployed on the top of the NFV MANO. Thus, the NFV promises to deliver flexibility to scale up/down network services according to the customer demand, QoE/QoS requirement of emerging Internet applications, and business-level policies. Furthermore, the NFV reduces capital expenditure (CAPEX) and operational expenditure (OPEX) costs by deploying NF on low-cost virtual infrastructure [39]. As mentioned in [4, 39], NFV introduces three major differences in how network services are provisioned as compared to previous practice: (i) decoupling of software from hardware platform, (ii) dynamic network operation and service provisioning, and (iii) greater flexibility for network functions deployment.

Table 6.2 A summary of NFV open source platforms, projects, and implementations.

Name	Technology	Company/ organization	Objective/functionality
OpenMANO	SDN and NFV	Telefonica	To provide a practical implementation of the NFV MANO reference architecture.
OSM	SDN and NFV	ETSI NFV	To support for a model-driven environment with data models aligned with ETSI NFV MANO
OPNFV	NFV	Linux Foundation	To facilitate the development of NFV components across various open source ecosystems.
ECOMP	SDN, NFV and Cloud	AT & T	To provide an enhanced control, orchestration and life cycle management of VNFs and the cloud platform where the VFs reside on.
Broadcom OpenNFV	NFV	Broadcom Corp.	To provide scalability, workload flexibility, and interoperability for the successful implementation of NFV.
OpenBaton [37]	NFV, Cloud	FOCUS	To develop a customizable framework capable of orchestrating network services across heterogeneous NFV infrastructures.
Cloudify	NFV, Cloud	Gigaspaces	A management and orchestration platform, designed to automate and deploy network services across multiple clouds, data centers, and stack environments.
ZOOM	NFV	TM Forum	To develop a platform for monitoring and optimization of Network Functions-as-a-Service (NFaaS).
T-NOVA	SDN and NFV	European Union	To develop a MANO platform for.
CloudNFV	SDN, NFV and Cloud	European Union	To provide an open management architecture that enables the NFV deployment in a cloud environment.
HP OpenNFV	NFV	European Union	To develop a NFV-architecture that allocates resources from an appropriate pool based on global resource management policies.

Table 6.2 (*Continued*)

Name	Technology	Company/ organization	Objective/functionality
Overture vSE	NFV	Overture Networks	To provide a carrier Ethernet server platform that enables SPs to host VNFs at the service edge.
Cisco ONS	SDN, NFV and Cloud	Cisco Systems, Inc.	To provide an Evolved Programmable Network (EPN) as a holistic systems-based approach for multivendor and multitenant data centers management.
F5 SDAS	SDN	F5 Synthesis	To provide a highly flexible application service platform for supporting the control and data plane programmability using interfaces such as *iControl* and *iRules*, respectively.

6.2.5 NFV Management and Orchestration (NFV MANO) Framework

The ETSI was the first to investigate the NFV idea in operator infrastructures [40], primarily to address the issues of providing flexible and agile services and to create a foundation for future network monetization. The NFV reference architecture [41, 42] shown in Figure 6.3 which includes two SDN controllers was introduced along with a proof of concept (PoC) [43]. The ETSI MANO framework consists of the following components, namely Network Function Virtualization Orchestrator (NFVO), Network Function Virtualization Infrastructure (NFVI), Network Management System (NMS), and VNFs and Services [4, 41]. The NFVO manages the physical and software resources that enable the virtualized infrastructure through orchestration and lifecycle management. The NFVO also manages global resources, creates network services, validates them, and authorizes NFVI resource requests. The NFVI creates an infrastructure consisting of both virtual (abstractions of the computing, storage, and network resources achieved through a virtualization layer) and physical resources (computing hardware for storage and network resources). Within the infrastructure domain of the operator, the VIM supervises and manages NFVI physical and virtual resources (vCompute, vStorage, and vNetwork resources). A VNFM can manage a single or multiple VNF instances of the same or different types. As indicated in Figure 6.3, the infrastructure SDN controller enables the needed connectivity for communicating the VNFs and its components while managing the underlying networking resources [44].

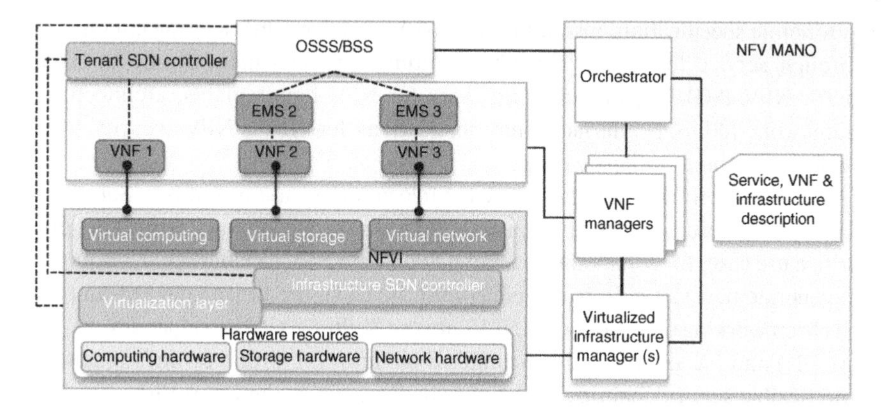

Figure 6.3 An integration of SDN controllers into the ETSI NFV reference architecture at the two levels required to achieve network slicing.

The tenant SDN controller manages dynamically the network resources in the forwarding plane which are used by tenants.

6.2.6 NFV Use Cases, Application Scenarios, and Implementation

Many NFV use cases and applications from academia and industry have been devised and deployed by telecomm operators and service providers (SPs). The majority of these use cases from the academia are based on ETSI guidelines [45] including deep packet inspection [46], RAN [47–49], customer premises equipment (CPE) [50–52], and evolved packet core (EPC) [53–55]. The key products and implementations from the industry include the following: CloudNFV [56], Huawei NFV Open Lab [57], HP OpenNFV [58], Intel Open Network Platform (Intel ONP) [59], Cisco Open Network Strategy [60], Alcatel-Lucent CloudBand [61, 62], and Broadcom Open NFV [63]. It is worth noting that all of the existing NFV implementations and platforms are focusing on open source and orchestration of operator's end-to-end service over NFVI supported by SDN and cloud technologies.

6.2.7 NFV Standardization Activities

Current NFV standardization activities span across many domains. The Network Function Virtualization Research Group (NFVRG) [64] of the Internet Research Task Force (IRTF) is active in developing new virtualized architectures with capabilities to provide support for NFs. The Internet Engineering Task Force Service Function Chaining Working Group (IETF SFC WG) [65] has put efforts to develop SFC architectural building blocks that address mechanisms for autonomic specification, instantiation of NFV instances, and steering data traffic through service functions. The ETSI's Industry Specification Group for NFV (ETSI NFV ISG) [66] provides use cases for NFV requirements, architectural framework [40, 67], interfaces and abstractions for NFVI, NFV security [68], performance, and resiliency [69].

As a follow-on activity, the Alliance for Telecommunications Industry Solutions (ATIS) NFV Forum (ATIS NFV Forum) [70] is devoted to identify, define, and prioritize use cases for collaboration among SPs where NFV capabilities are required to generate new value. In partnership with the ETSI, the Broadband Forum [71] has been working on introducing NFV into the Multi-Service Broadband Network (MSBN) and has established a virtualized platform for virtual business gateway and flexible service chaining [72]. Other bodies such as the Open Virtualization Format (OVF) and the ITU-T SG13 have been working toward defining the functional requirements and architecture of the network virtualization for NGN as well as the portability and deployment of both virtual and physical machines across multiple platforms. Along with the above standardization activities, many collaborative projects pushing the NFV implementations include the Open Platform for Network Function Virtualization (OPNFV) [73], the Zero-time Orchestration, Operations and Management (ZOOM) [74] and OpenMANO [75], Unifying compute and network virtualization (UNIFY) [76, 77], among others.

6.2.8 Network Hypervisors, Containers, and Virtual Machines

A network hypervisor is a software that creates and runs VMs. Hypervisors [78] are network elements that abstract physical infrastructure (e.g. control functions and communication channels) into conceptually isolated virtual network slices. In SDN, the main functions that are implemented by a hypervisor to create virtual SDN network are the following: (i) network attribute virtualization, (ii) abstraction and management of the physical SDN network, where isolated virtual SDN networks can be created and controlled by the respective virtual SDN controllers. Various hypervisors [79] have been implemented including the following: OpenSlice, RadioVisor, FlowVisor [80], OpenVirteX [81], MobileVisor [4], etc. The network hypervisors are comprehensively discussed in [79]. The network hypervisors such as OpenVirteX and FlowVisor are designed focusing on a fixed and wired SDN network slicing.

A VM is a software program that emulates the functionality of a physical hardware or computing system. It runs on top of the hypervisor, which replicates the functionality of the underlying physical hardware resources [4]. These resources may be referred to as the host machine, while the VM that runs on the hypervisor is often called a guest machine. The VM contains all necessary elements to run the apps, including computing, storage, memory, networking,

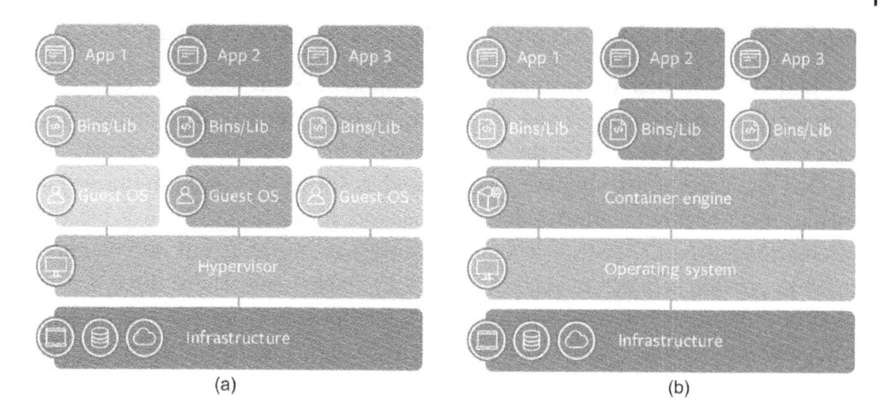

Figure 6.4 A comparison between containers and virtual machines. (a) Virtual machines. (b) Containers.

and a hardware functionality available as a virtualized system. The VM may also contain the necessary system binaries and libraries to run the apps. The actual operating system (OS), however, is managed and executed using the hypervisor [82]. Containerization creates abstraction at an OS level that allows individual, modular, and distinct functionality of the app to run independently. As a result, several isolated workloads, the containers – can dynamically operate using the same physical resources. Containers can run (i) on top bare metal servers, (ii) on top of hypervisors, (iii) in cloud infrastructure. Figure 6.4 shows a comparison between a VM (running on a hypervisor) and containers that run on.

Containers share all necessary capabilities with the VM to operate as an isolated OS environment for a modular app functionality with one key difference. Using a containerization engine, such as the Docker Engine, containers create several isolated OS environments within the same host system kernel, which can be shared with other containers dedicated to run different functions of the app. Only bins, libraries, and other runtime components are developed or executed separately for each container, which makes them more resource efficient as compared to VMs. Containers are particularly useful in developing, deploying, and testing modern distributed apps and microservices that can operate in isolated execution environments on same host machines. Containerization eliminates the need for developers to write application code into many VMs running distinct app components in order to retrieve compute, storage, and networking resources. A fully functional application component can be run in its own isolated environment without influencing other app components or software. During execution, there are no conflicts between libraries or app components, and the application container can easily move between cloud and data center instances.

6.2.9 Multiaccess Edge Computing (MEC): From Clouds to Edges

CC [83] is the on-demand availability of computer system resources, especially data storage (cloud storage) and computing power, without direct active management by the user. Today, the key roles of a SP on a CC environment are divided into two parts: (i) infrastructure providers (InPs) who perform the overall management of cloud platforms and lease network resources using a usage-based pricing model; and (ii) SPs – who rent network resources (from one or many InPs) to provide services to customers. The CC model consists of three service models (Infrastructure-as-a Service (IaaS), Platform-as-a-Service (PaaS), and Software-as-a-Service (SaaS)) as described in [4, 82]. Figure 6.5 illustrates the three service models and their mapping to the NFV-MANO reference architecture. The IaaS such as the Amazon Web Services (AWS), Microsoft Azure, and Google Compute Engine (GCE) enable users to access, monitor, and manage distant data-center infrastructures (e.g. compute, storage, and networking services) using self-service models, while the PaaS offers a platform where users can develop, run, and administer various applications without having to worry about the complexities of setting up and managing cloud infrastructure. The SaaS provides environment where users can use applications and services running on a cloud infrastructure.

MEC moves the computing of traffic and services from a centralized cloud to the edge of the network and closer to the customer. Collecting and processing data closer to the customer reduces latency and brings real-time performance to high-bandwidth applications such as video streaming. MEC provides CC capabilities to satisfy the high-demanding requirements of 5G such as throughput and an improved QoE for the end users [84]. The MEC platform enables distributed edge computing by processing content at the edge using either a server or a CPE. Applications at the edge require a high bandwidth and low-latency environment. To achieve that service, providers create distributed data centers or distributed

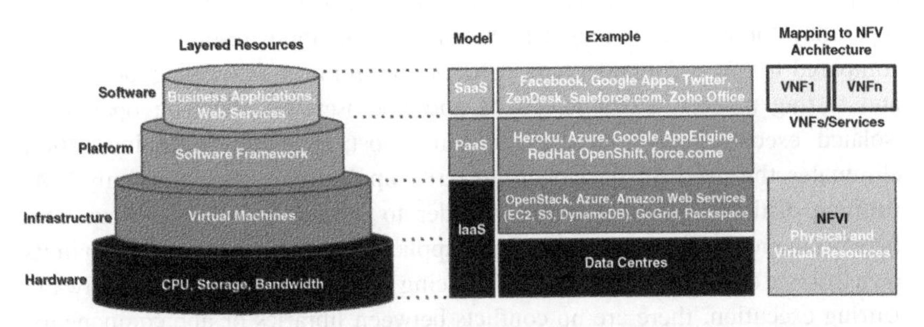

Figure 6.5 Cloud computing service models and their mapping to part of the NFV reference architecture. Source: Mijumbi et al. [82].

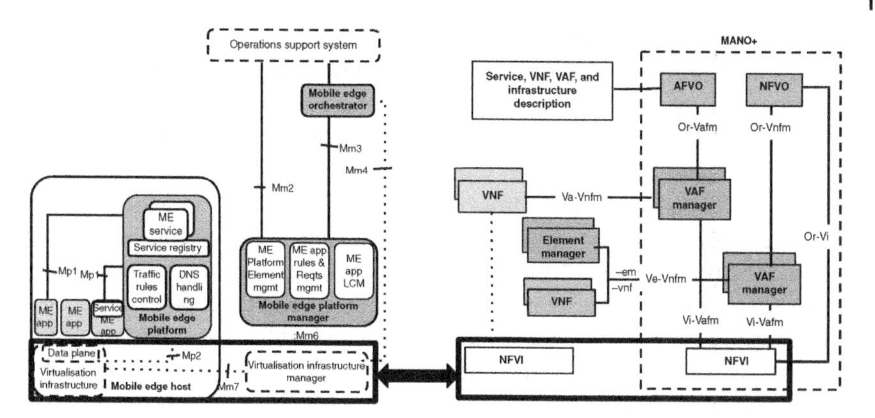

Figure 6.6 Multiaccess edge computing architecture and its mapping to NFV MANO.

clouds. A software-defined access layer, for example, could also be used as an extension of a distributed cloud. Most edge computing initiatives are being developed using open source hardware and software that leverage cloud and virtualization paradigms, including SDN and NFV. As shown in Figure 6.6, SDN and NFV are to play a key role in the 5G MEC architecture to provide a virtual environment (e.g. vCompute, vStorage, and networking resources) for MEC applications [85]. The MEC architecture consists of a MEC platform that forms a central point where servers are hosted. The MEC data plane consists of traffic offload functions (TOFs) which provide interfaces with policy-based packet monitoring to both MEC Apps and MEC services.

The MEC server, which is close to the users, ensures that bandwidth-intensive services such as video streaming have a low latency. Through caching contents at the MEC server, a similar concept to ICN [86], MEC in 5G reduces the volume of data transmitted at the 5G core network for processing, enabling real time, and application flow information as well as efficient use of available resources. MEC opens-up new significant opportunities and capabilities for application/ SPs and network operators to deploy disruptive and innovative 5G services in the area of IoT, VR/AR, immersive experience, and big data analytics. Since 5G applications and services are placed at the edge, response to requests can be improved to provide an enhanced QoE for the end users.

6.3 Conclusion

The concepts of network softwarization, as well as their operational principles, are presented in this chapter. We use promising technologies such as SDN, NFV, and MEC to add softwarization methods in mobile edge networks, core networks, and transport networks. 5G network softwarization is set to facilitate future network management and orchestration of resources from SPs to the end users. Identifying how network services and associated resources that are implemented according to an SDN architecture might be incorporated within the NFV architectural framework has been an achievable step so far in the design patterns of network softwarization. NFV decouples NFs from dedicated hardware devices, whereas SDN decouples the control plane from the data/packet forwarding plane. Despite the fact that the two (SDN and NFV) have a lot in common, the fundamental difference is that SDN necessitates a new network platform with decoupled control and data forwarding planes. The chapter also provides a highlight on the relationship among VMs, CC, and NFV. The output of all of these technologies toward making 5G network slicing a reality, as expected and advocated by vendors, operators, and SPs, has to be observed from both academics and industry.

Bibliography

1 Lake, D., Wang, N., Tafazolli, R., and Samuel, L. (2021). Softwarization of 5G networks-implications to open platforms and standardizations. *IEEE ACCESS* 9: 88902–88930.

2 Afolabi, I., Taleb, T., Samdanis, K. et al. (2018). Network slicing and softwarization: a survey on principles, enabling technologies and solutions. *IEEE Communication Surveys and Tutorials* 20 1–24.

3 Condoluci, M., Sardis, F., and Mahmoodi, T. (2016). Softwarization and virtualization in 5G networks for smart cities. *International Internet of Things Summit* 179–186.

4 Barakabitze, A.A., Ahmad, A., Mijumbi, R., and Hines, A. (2020). 5G network slicing using SDN and NFV: a survey of taxonomy, architectures and future challenges. *Computer Networks* 167: 1–40.

5 Sanchezl, J., Yahial, I.G.B., Crespi, N. et al. (2014). Softwarized 5G networks resiliency with self-healing. *2014 1st International Conference on 5G for Ubiquitous Connectivity (5GU)*, pp. 229–233, November 2014.

6 Ravindran, R., Chakraborti, A., Amin, S.O. et al. (2017). 5G-ICN: Delivering ICN services over 5G using network slicing. *IEEE Communications Magazine* 55 (5): 101–107.

7 Tran, T.X., Hajisami, A., Pandey, P., and Pompili, D. (2017). Collaborative mobile edge computing in 5G networks: new paradigms, scenarios, and challenges. *IEEE Communications Magazine* 55 (4): 54–61.

8 Liu, Y., Fieldsend, J.E., and Min, G. (2017). A framework of fog computing: architecture, challenges, and optimization. *IEEE Access: Special Section on Cyber-Physical Social Computing and Networking* 5: 25445–25454.

9 View on 5G Architecture: 5G PPP Architecture Working Group. https://5g-ppp.eu/wp-content/uploads/2014/02/5G-PPP-5G-Architecture-WP-For-public-consultation.pdf (accessed 20 March 2019).

10 Lara, A., Kolasani, A., and Ramamurthy, B. (2014). Network innovation using OpenFlow: a survey. *IEEE Communication Surveys and Tutorials* 16 (1): 493–512.

11 Doria, A., Salim, J.H., Haas, R. et al. (2010). Internet Engineering Task Force (IETF): Forwarding and Control Element Separation (ForCES).

12 Smith, M., Dvorkin, M., Laribi, Y. et al. (2014). OpFlex Control Protocol," Internet Draft, Internet Engineering Task Force. http://tools.ietf.org/html/draft-smith-opflex-00 (accessed 06 March 2018).

13 Taylor, M. (2014). A guide to NFV and SDN. White paper," Metaswitch Networking. London, UK. http://www.metaswitch.com/sit,es/default/files/Metaswitch.WhitePaper.NFVSDN.final.rs.pdf (accessed 06 March 2018).

14 vmware (2015). The VMware NSX Network Virtualization Platform. Technical white paper, VMware, Palo Alto, CA, USA. https://www.vmware.com/content/dam/digitalmarketing/vmware/en/pdf/whitepaper/products/nsx/vmware-nsx-network-virtualization-platform-white-paper.pdf (accessed 06 March 2018).

15 Floodlight (2012). Floodlight is a Java-based OpenFlow controller. http://floodlight.openflowhub.org/ (accessed 06 March 2018).

16 Erickson, D. (2013). The Beacon OpenFlow controller. *in Proceedings of the 2nd ACM SIGCOMM Workshop on Hot Topics in Software Defined Networking, ser. HotSDN*, pp. 13–18, August 2013.

17 Nippon Telegraph and Telephone Corporation (2012). Ryu Network Operating System. http://osrg.github.com/ryu/ (accessed 06 March 2018).

18 Tootoonchian, A. and Ganjali, Y. (2010). HyperFlow: A distributed control plane for OpenFlow. *Proceedings of the 2010 Internet Network Management Conference on Research on Enterprise Networking*, ser. INM/WREN'10, pp. 1–6, August 2010.

19 Hp SDN controller architecture, Hewlett-Packard Development Company, L.P, Tech. Rep., September 2013. http://h17007.www1.hpe.com/docs/networking/solutions/sdn/devcenter/06.-.HP.SDN.Controller.Architecture.TSG.v1.3013-10-01.pdf (accessed 06 March 2018).

20 Phemius, K., Bouet, M., and Leguay, J. (2013). DISCO: Distributed Multidomain SDN Controllers, pp. 1–8. *ArXiv e-prints.*

21 Berde, P., Gerola, M., Hart, J. et al. (2014). ONOS: Towards an open, distributed SDN OS. *in Proceedings of the 3rd Workshop on Hot Topics in Software Defined Networking*, pp. 1–6, August 2014.

22 ONF TR-526 (2016). Applying SDN Architecture to 5G Slicing.

23 Ordonez-Lucena, J., Ameigeiras, P., Lopez, D. et al. (2017). Network slicing for 5G with SDN/NFV: concepts, architectures, and challenges. *IEEE Communications Magazine* 55 (5): 80–87.

24 Devlic, A., Hamidian, A., Liang, D. et al. (2018). NESMO: Network slicing management and orchestration framework. *IEEE International Conference on Communications Workshops*, May 2018.

25 Apache CloudStack. http://cloudstack.apache.org (accessed 06 March 2018).

26 OpenStack. http://www.openstack.org/ (accessed 06 March 2018).

27 OpenDayLight. http://www.opendaylight.org (accessed 06 March 2018).

28 Metro Ethernet Forum (MEF). http://metroethernetforum.org (accessed 06 March 2018).

29 Cascone, C., Pollini, L., Sanvito, D., and Capone, A. (2015). Traffic management applications for stateful SDN data plane. *4th European Workshop on Software Defined Networks*, November 2015.

30 Yeganeh, S.H., Tootoonchian, A., and Ganjali, Y. (2013). On scalability of software-defined networking. *IEEE Communications Magazine* 51 (2): 136–141.

31 Bosshart, P., Daly, D., Gibb, G. et al. (2014). P4: Programming protocol-independent packet processors. *ACM SIGCOMM Computer Communication Review* 44 (3): 87–95.

32 Bianchi, G., Bonola, M., Capone, A., and Cascone, C. (2014). OpenState: Programming platform-independent stateful OpenFlow applications inside the switch. *ACM SIGCOMM Computer Communication Review* 44 (2): 44–51.

33 Arashloo, M.T., Koral, Y., Greenberg, M. et al. (2015). SDPA: Enhancing stateful forwarding for software-defined networking. *IEEE 23rd International Conference on Network Protocols*, pp. 323–333, November 2015.

34 Song, H. (2013). Protocol-oblivious forwarding: unleash the power of SDN through a future-proof forwarding plane. *SIGCOMM HotSDN*, pages 127–132, August 2013.

35 Arashloo, M.T., Koral, Y., Greenberg, M. et al. (2016). SNAP: Stateful network-wide abstractions for packet processing. *ACM SIGCOMM Conference*, pp. 29–43, August 2016.

36 Zhang, X., Cui, L., Wei, K. et al. (2021). A survey on stateful data plane in software defined networks. *Computer Networks* 184. 1–24

37 Carella, G.A., Pauls, M., Magedanz, T. et al. (2017). Prototyping NFV-based multi-access edge computing in 5G ready networks with open baton. *IEEE Conference on Network Softwarization (NetSoft)*, July 2017.

38 Barakabitze, A.A., Barman, N., Ahmad, A. et al. (2020). QoE management of multimedia services in future networks: a tutorial and survey. *IEEE Communication Surveys and Tutorials*. 22 (1): 526–565.

39 Mijumbi, R., Serrat, J., Gorricho, J.L. et al. (2016). Management and orchestration challenges in network functions virtualization. *IEEE Communications Magazine* 54 (1): 98–105.

40 ETSI (2013). ETSI GS NFV V1.1.1: Network Functions Virtualisation (NFV); Architectural Framework. http://www.etsi.org/deliver/etsi.gs/nfv/001.099/002/01.01.01.60/gs.nfv002v010101p.pdf (accessed 06 March 2018).

41 Barakabitze, A.A., Barman, N., Ahmad, A. et al. (2019). QoE management of multimedia streaming services in future networks: a tutorial and survey. *IEEE Communication Surveys and Tutorials* 22 (1): 526–565.

42 ETSI (2015). Network Functions Virtualisation (NFV); Ecosystem; Report on SDN Usage in NFV Architectural Framework. http://www.etsi.org/deliver/etsi.gs/NFV-EVE/001.099/005/01.01.01.60/gs.NFV-EVE005v010101p.pdf (accessed 06 March 2018).

43 ETSI (2014). ETSI GS NFV-PER 002 V1.1.2: Network Functions Virtualisation (NFV); Proof of Concepts; Framework. http://www.etsi.org/deliver/etsi.gs/NFV-PER/001.099/002/01.01.02.60/gs.NFV-PER002v010102p.pdf (accessed 06 March 2018).

44 ETSI GS NFV-INF 001 (2015). Network Functions Virtualisation (NFV); Infrastructure Overview, v. 1.1.1.

45 ETSI (2013). ETSI GS NFV 001 V1.1.1: Network function virtualization. Use cases, ETSI Ind. Spec. Group (ISG) Netw. Functions Virtualisation (NFV). http://www.etsi.org/deliver/etsi.gs/nfv/001.099/001/01.01.01.60/gs.nfv001v010101p.pdf (accessed 06 March 2018).

46 Bremler-Barr, A., Harchol, Y., Hay, D., and Koral, Y. (2014). Deep packet inspection as a service. *in Proceedings of the 10th ACM International CoNEXT*, pp. 271–282.

47 Sabella, D., Rost, P., Sheng, Y. et al. (2013). RAN as a service: challenges of designing a flexible RAN architecture in a cloud-based heterogeneous mobile network. *IEEE Future Network and Mobile Summit (FutureNetworkSummit)*, pp. 1–8.

48 Chih-Lin, I., Huang, J., Duan, R. et al. (2014). Recent progress on C-RAN centralization and cloudification. *EEE Access* 2: 1030–1039.

49 Wang, R., Hu, H., and Yang, X. (2014). Potentials and challenges of C-RAN supporting multi-RATs toward 5G mobile networks. *EEE Access* 2: 1187–1195.

50 Aleksic, S. and Miladinovic, I. (2014). Network virtualization: paving the way to carrier clouds. *in Proceedings of the IEEE 16th International Telecommunications Network Strategy and Planning Symposium*, pp. 1–6, September 2014.

51 Vilalta, R., Munoz, R., Mayoral, A. et al. (2015). Transport network function virtualization. *Journal of Lightwave Technology* 33 (8): 1557–1564.

52 Vilalta, R., Munoz, R., Casellas, R. et al. (2014). Transport PCE network function virtualization. *in Proceedings of the IEEE ECOC*, pp. 1–3, September 2014.

53 Sama, M.R., Contreras, L.M. et al. (2015). Software-defined control of the virtualized mobile packet core. *IEEE Communication Magazine* 53 (2): 107–115.

54 Gebert, S., Hock, D., Zinner, T. et al. (2014). Demonstrating the optimal placement of virtualized cellular network functions in case of large crowd events. *in Proceedings of the ACM Conference SIGCOMM*, pp. 359–360, August 2014.

55 Vo, N.-S., Duong, T.Q., Tuan, H.D., and Kortun, A. (2017). Optimal video streaming in dense 5G networks with D2D communications. *IEEE Access* 6: 209–223.

56 CloudNFV (2015). http://www.cloudnfv.com/ (accessed 06 March 2018).

57 Huawei NFV Open Lab (2015). http://pr.huawei.com/en/news/hw-411889-nfv.htm#.WkPoloZl.IU (accessed 06 March 2018).

58 HP (2015). HP OpenNFV Reference Architecture. http://www8.hp.com/us/en/cloud/nfv-overview.html? (accessed 06 March 2018).

59 Intel. Intel Open Network Platform. https://www.intel.com/content/www/us/en/communications/intel-open-network-platform.html (accessed 06 March 2018).

60 Cisco (2014). Cisco NFV solution: enabling rapid service innovation in the era of virtualization, White Paper. https://www.cisco.com/c/dam/global/shared/assets/pdf/sp04/nfv-solution.pdf (accessed 06 March 2018).

61 Alcatel-Lucent's ClouBand. http://www.alcatel-lucent.com/solutions/cloudband, note =.

62 Nokia (2014). CloudBand with OpenStack as NFV platform. Strategic white paper, NFV insights series, Alcatel Lucent RedHat, Boulogne Billancourt, France, Tech. Rep. www.alcatel-lucent.com/ (accessed 12 August 2022).

63 Broadcom Open NFV (2015). http://www.broadcom.com/press/release.php?id=s827048 (accessed 06 March 2018).

64 Internet Research Task Force (2015). Network Function Virtualization Research Group (NFVRG). https://irtf.org/nfvrg (accessed 06 March 2018).

65 Datatracker (2015). The Internet Engineering Task Force (IETF) Service Function Chaining (SFC) Working Group (WG). https://datatracker.ietf.org/wg/sfc/charter/ (accessed 12 August 2022).

66 ETSI (2014). ETSI GS NFV-SWA 001: ETSI Industry Specification Group (ISG) NFV. Network Functions Virtualisation (NFV); Virtual Network Functions Architecture. http://www.etsi.org/deliver/etsi.gs/NFV-SWA/001.099/001/01.01.01.60/gs.NFV-SWA001v010101p.pdf (accessed 06 March 2018).

67 ETSI (2014). ETSI GS NFV 003 V1.2.1: Network Functions Virtualisation (NFV); Terminology for main concepts in NFV, ETSI Ind. Spec. Group (ISG) Netw. Functions Virtualisation (NFV). http://www.etsi.org/deliver/etsi.gs/NFV-

MAN/001.099/001/01.01.01.60/gs.nfv-man001v010101p.pdf (accessed 06 March 2018).

68 ETSI (2014). ETSI GS NFV-SEC 003 V1.1.1: Network Functions Virtualisation (NFV); NFV security; Security and trust guidance. ETSI Ind. Spec. Group (ISG) Netw. Functions Virtualisation (NFV).

69 ETSI (2015). ETSI GS NFV-REL 001 V1.1.1: Network Functions Virtualisation (NFV); Resiliency requirements," ETSI Ind. Spec. Group (ISG) Netw. Functions Virtualisation (NFV).

70 Atis (2015). Alliance for Telecommunications Industry Solutions, network functions virtualization forum, ATIS Netw. Functions Virtualization (NFV). http://www.atis.org/NFV/index.asp (accessed 06 March 2018).

71 The Broadband Forum (2015). https://www.broadband-forum.org/ (accessed 12 August 2022).

72 Broadband Forum (2015). The Broadband Forum Technical Work in Progress. https://www.broadband-forum.org/standards-and-software/technical-specifications/technical-reports (accessed 12 August 2022).

73 Opnfv (2015). Open Platform for NFV (OPNFV). https://www.opnfv.org/about (accessed 12 August 2022).

74 Zero-time Orchestration, Operations and Management(ZOOM), Tele-Manage. Forum. *Tech. Rep*, August 2014.

75 Lopez, D.R. (2015). OpenMANO: The Data plane ready open source NFV MANO stack. *Proceedings of the IETF Meet*, pp. 1–28.

76 Sonkoly, B., Szabo, R., Jocha, D. et al. (2015). UNIFYing cloud and carrier network resources: an architectural view. *Proceedings of the 2015 IEEE Global Communications Conference(GLOBECOM)*, pp. 1–7, December 2015.

77 Szabo, R., Kind, M., Westphal, F.-J. et al. (2015). Elastic network functions: opportunities and challenges. *IEEE Network* 29 (3): 15–21.

78 Huang, S., Griffioen, J., and Calvert, K.L. (2014). Network hypervisors: enhancing SDN infrastructure. *Computer Communications* 46: 87–96.

79 Blenk, A., Basta, A., Reisslein, M., and Kellerer, W. (2016). Survey on network virtualization hypervisors for software defined networking. *IEEE Communication Surveys and Tutorials* 18 (1): 655–685.

80 Sherwood, R., Gibb, G., Yap, K.-K. et al. (2009). SFlowVisor: A network virtualization layer. *OpenFlow Consortium*. 26 1–14.

81 Al-Shabibi, A., De Leenheer, M., Koshibe, A. et al. (2014). OpenVirteX: Make your virtual SDNs programmable. *Proceedings of the 3rd Workshop on Hot Topics in Software Defined Networks (HotSDN)*, pp. 25–30, August 2014.

82 Mijumbi, R., Serrat, J., Gorricho, J.-L. et al. (2016). Network function virtualization: state-of-the-art and research challenges. *IEEE Communication Surveys and Tutorials* 18 (1): 236–262.

83 Sanaei, Z., Abolfazli, S., Gani, A., and Buyya, R. (2014). Heterogeneity in mobile cloud computing: taxonomy and open challenges. *IEEE Communication Surveys and Tutorials* 16 (1): 369–392.

84 Tselios, C. and Tsolis, G. (2016). On QoE-awareness through virtualized probes in 5G networks. *2016 IEEE 21st International Workshop on Computer Aided Modelling and Design of Communication Links and Networks (CAMAD)*, pp. 159–164, October 2016.

85 Pham, Q.-V., Fang, F., Ha, V.N. et al. (2020). A survey of multi-access edge computing in 5G and beyond: fundamentals, technology integration, and state-of-the-art. *IEEE Access* 8: 116974–117017.

86 Barakabitze, A.A., Sun, L., Mkwawa, I.-H., and Ifeachor, E. (2016). A novel QoE-aware SDN-enabled, NFV-based management architecture for future multimedia applications on 5G systems. *8th International Conference on Quality of Multimedia Experience (QoMEX)*, June 2016.

7

Management of Multimedia Services in Emerging Architectures Using Big Data Analytics: MEC, ICN, and Fog/Cloud computing

Network softwarization technologies that leverage Network Function Virtualization (NFV) and Software-Defined Network (SDN) have demonstrated to be efficient in terms of managing, control, and monitoring the end-users Quality of Experience (QoE) as discussed in Chapters 5 and 6. With this advancement, there are other emerging network architectures that can used for providing interactive and immersive video services to end-users while taking the advantages of SDN and NFV. Emerging architectures along with the integration of SDN and NFV that can help to cope with the growing multimedia streaming services over the Internet include: Multi-Access Edge Computing (MEC), fog/cloud and Information/Content-Centric Networking (ICN), self-driving S3D network architecture and Artificial Intelligence/Machine Learning (AI/ML) -data driven network management architectures. This chapter presents emerging QoE-aware/driven architectures for adaptive video streaming.

7.1 QoE-Aware/Driven Adaptive Streaming Over MEC Architectures

The environment of MEC offers low latency to applications with real-time insight, high bandwidth, radio network information, and location awareness. Recent architectural attempts include the QoE-oriented approach [1] that supports streaming services provisioning to end users dynamically in SDN-NFV-enabled environment. The authors in [2] introduce a QoE monitoring probe that can adapt to the end users' QoE demand and services delivery characteristics based on the knowledge from the cell topology, Radio Access Network (RAN) type and network resources allocation. The performance analysis and evaluation of a MEC-based adaptive videos streaming services in future networked NFV-enabled platforms is proposed in [3] where a context-aware adaptive video prefetching (MVP) architecture that can provide QoE-assured 4K video on demand delivery

Multimedia Streaming in SDN/NFV and 5G Networks: Machine Learning for Managing Big Data Streaming, First Edition. Alcardo Barakabitze and Andrew Hines.

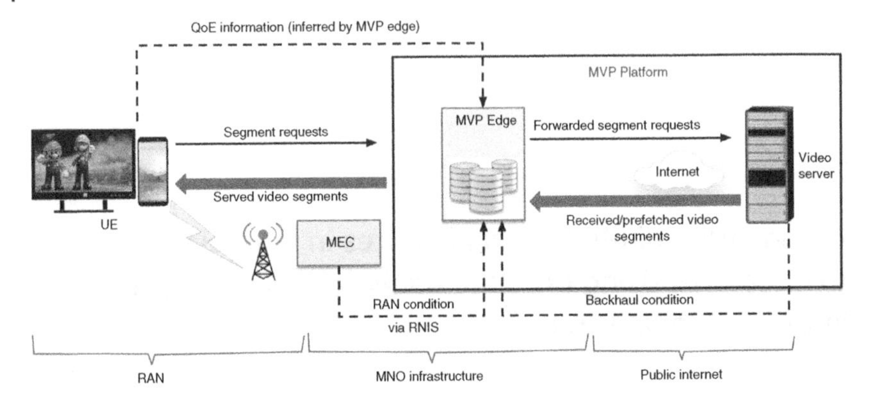

Figure 7.1 Mobile edge virtualization with adaptive prefetching. Source: Alexas_Fotos/Pixabay, ICCCC/Wikipedia/CC0 1.0.

to the end-users in mobile edge environment is introduced in [4]. The proposed MVP architecture is indicated in Figure 7.1 where the video segment requests from users during Video on Demand (VoD) sessions is managed by the MVP edge component. The MVP edge is an intelligent component that is aware of the streaming characteristics of every UE. It also has knowledge of the end-users' QoE Influencing Factors (IFs) such as buffer status of streaming sessions and the throughput history of all video segments that are downloaded. It is worth noting that, MVP edge disseminates periodically the RAN contest information in real-time via the Radio Network Information Services (RNIS) module in MNO-owned MEC servers.

The MVP edge servers that have minimum access latency can provide a video segment based on the requests from the end-user. When the video segment can not be offered by the closest edge server, then the request is forwarded by the MVP edge to the original video source. The MVP edge also performs network traffic monitoring, optimization, and adaptation of video quality based on the real-time streaming information received from the UEs and the network. Video content providers can make use of this approach and perform adaptive prefetching for each VoD session using embedded content intelligence at the MVP edge where video segments are cached, prefetched, and downloaded in advance (e.g. during the UE's request progress). Several attempts have been made including a demonstration of edge computing architecture that can performs QoE estimation in real-time for Dynamic Adaptive Streaming over HTTP (DASH)-based mobile video applications [5]. The authors in [6] proposes a MEC -based architecture to improve the QoE of video delivery service in urban spaces while Li et al. [7] propose a MEC-assisted video delivery approach that consider the number of active UE and the available network bandwidth to maximize the QoE-fairness for all DASH clients

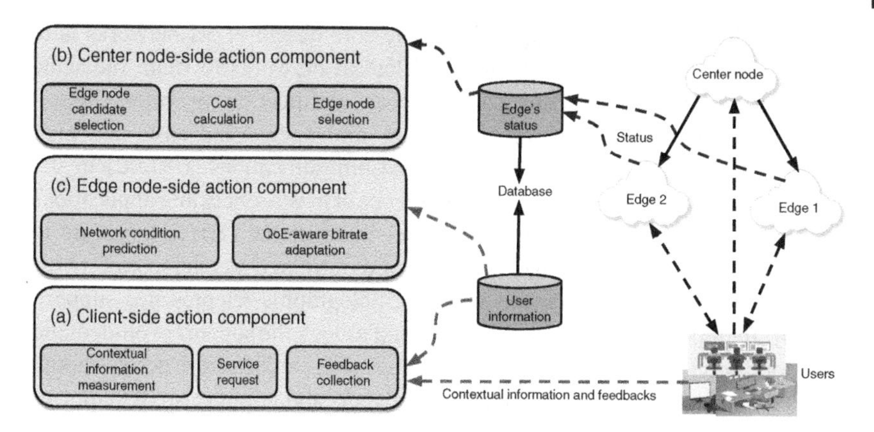

Figure 7.2 An overview of QoE-aware control plane for adaptive streaming service in MEC.

in heterogeneous wireless networks. Moreover, an Edge-based Transient Holding of Live Segment (ETHLE) architecture is introduced by Ge et al. [8] to provide seamless 4K live video streaming experience the end-users. The buffer and initial startup delay and live video streaming latency are significantly reduced by this approach. The QoE driven mobile-edge caching placement optimization architecture that can achieve adaptive video streaming is introduced by the authors in [9]. The proposed approach considers different constraints on the network edge server (e.g. initial startup delay, the backhaul link, storage capacity) to reduce the video streaming distortion for DASH clients.

Figure 7.2 illustrates three components (client-side action, edge node-side action and center node-side action) of a QoE-aware Control plane for adaptive Streaming Service (QCSS) over MEC infrastructure [10]. The edge node-side action component maximizes the end users' QoE by performing QoE-aware bitrate adaption and network throughput prediction mechanisms while the client-side action component forwards the end-user's service requests to the MEC nodes and performs measurements on the contextual information (e.g. play status, a set of Quality of Services [QoS] attributes, and activity) collected as feedback from the UEs. However, moving computing and transmission loading from the center node prevents end-user clients from accessing video contents from the center node simultaneously, which may cause bandwidth bottlenecks [1].

7.2 QoE-aware Self-Driving 3D Network Architecture

Previous network generations including 5G networks were designed to provide better connectivity, increased link capacity, sufficient communication coverage, and edge/cloud computing support [11]. The core design of 5G networks is to allow multiple isolated and self-contained logical networks to share the same physical network infrastructure. 5G network architecture provides support to different services in terms of low latency and operational costs and high reliability (Ultra-Reliable and Low Latency Communications, URLLC) in the two-dimensional (2D) space [12]. The emergence of new services and applications in 6G and beyond networks requires a three-dimensional (3D) architecture to provide ubiquitous 3D coverage, seamless and extremely service connectivity, pervasive connectivity, unmanned mobility, holographic telepresence support, etc. [13]. To improve radio access capability and unlock the support of on-demand edge cloud services in 3D space, the 6G heterogeneous architecture design should complement and integrate with terrestrial networks and non-terrestrial platforms such as low-orbit satellites, balloons, and drones. To facilitate various 3D services and use cases in 3D space, future 6G 3D architecture should extend the concept of 5G network slicing and apply it across both terrestrial and non-terrestrial nodes. The 3D architecture in 6G networks should provide the management and orchestration of computing, communication and caching resources on demand basis, at any time and everywhere using AI/ML-based algorithms. Strinati et al. [14] propose a hierarchical 3D 6G network architecture shown in Figure 7.3 that unify diverse 3D network nodes which are distributed over terrestrial and non-terrestrial platforms. The Low Attitude Platforms (LAPs), High Altitude Platform (HAPs), aerial nodes (unmanned aerial vehicles [UAVs]) and Low Earth Orbit/Geostationary orbit (LEO/GEO) satellites are placed at various layers in the architecture.

The LAPs and HAPs offer high flexibility, mobility and adaptive coverage capacity for ground mobile users at low cost. The Integrated Access and Backhaul (IAB) nodes form 3D base stations and 3D relay to provide efficiency and support additional use cases such as Time Sensitive Communications (TSC) and intelligent transportation systems. The IAB supports QoS/QoE prioritization of the traffic on the backhaul link, flexible resource usage between access and backhaul link and topology adaptability in case of link failure. The 3D connectivity services shown in the architecture provide the needed flexibility to accommodate various applications and services including 3D intelligent services, interactive 3D video, real-time 3D traffic monitoring, and control and management [14]. The 3D network architecture supports the network slicing concept to support different vertical use cases and services provisioned in 3D space across both terrestrial nodes and non-terrestrial nodes. The AI-based cognitive decision making and intent-based End-to-End (E2E) service management will play a key role for network control and management by offering intelligent routing selection and load balancing approaches across 3D network layers. It should also provide 3D remote sensing mechanisms for pollution monitoring, land management

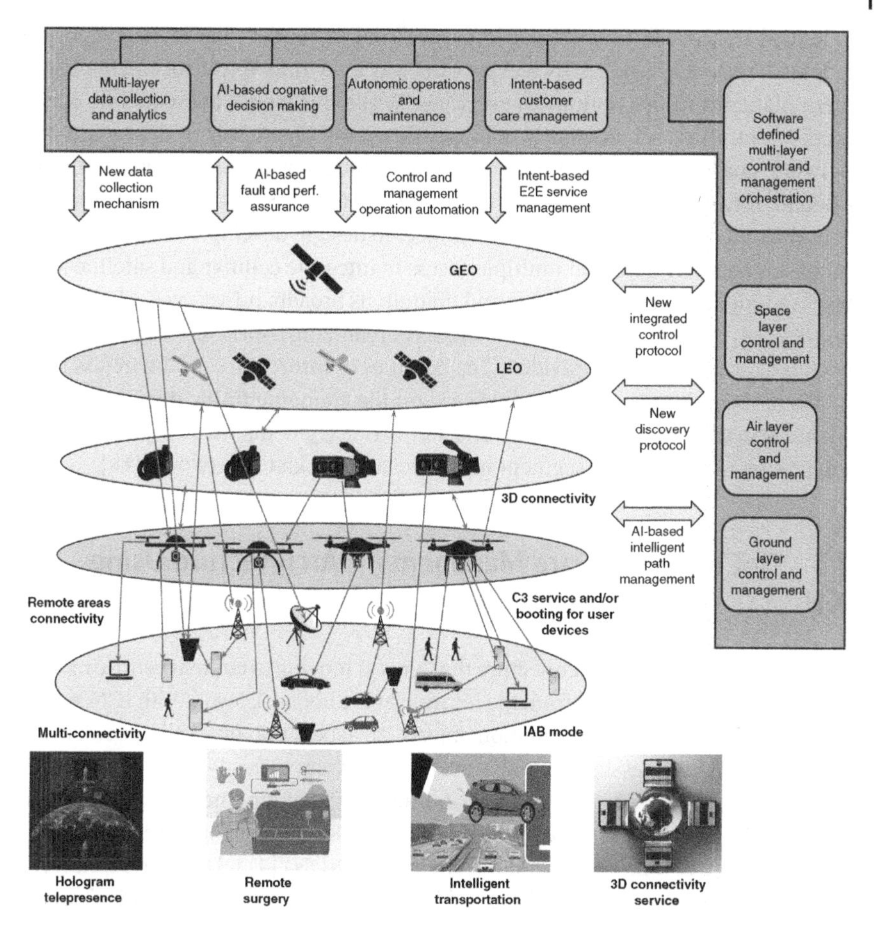

Figure 7.3 Hierarchical 3D network system architecture. Source: Strinati et al. [14] / with permission of John Wiley & Sons.

and agricultural services as well as meteorological measurements. It is worth mentioning that the 3D network architecture and 3D multi-connectivity will allow the user equipment to establish multiple different traffic links with 3D network nodes. This improves the service performance of UEs through dynamic load balancing strategies and intelligent control and management AI/ML-based algorithms [15].

The development of 3D-based 6G network architecture is taking place within several projects. The H2020 VIrtualized hybrid satellite-TerrestriAl systems for resilient and fLexible future networks (VITAL) project addresses the integration of terrestrial and satellite networks through two key innovations [16]: (i) to bring NFV concept into the satellite domain, and (ii) enable management of SDN-based federated resources in hybrid SatCom-terrestrial networks. The management of federated resources using SDN will pave the way for a unified control and management plane for future multimedia services. While this would allow mobile operators to optimize and efficiently manage the overall operations of the hybrid 3D network, enabling NFV into SatCom domain would allow them to offer virtualized satellite networks to third-party providers. The 5G-ALLSTAR project [17] leverages the outcomes of 5GCHAMPION project to design, develop, and evaluate the multi-connectivity based on multiple access to integrate cellular and satellite networks to support seamless reliable and ubiquitous broadband services. The H2020 SANSA [18] project proposes an efficient self-organizing hybrid terrestrial-satellite backhaul architecture to provide: (i) capabilities of future terrestrial wireless networks to reconfigure automatically based on the changing traffic demands, (ii) a shared spectrum between satellite and terrestrial segments, and (iii) a seamless integration of the satellite segment into terrestrial backhaul networks [18].

7.3 QoE-driven/aware Management Architecture Using ICN

ICN [38, 39] is a network architecture that aims at moving to content centric model where content is decoupled from the location at the network level. With ICN, naming is directly applied to information objects, an aspect that gives clients full control over streaming sessions and also over their devices to easily adapt to the multimedia network based on device capabilities and changing network conditions [11, 40]. The architectures that follows ICN principles include the Content-Centric Networking (CCN) or Named Data Networking (NDN) [41], Data-Oriented Network Architecture (DONA) [42], etc. Table 7.1 provides a summary of QoE-Centric management Strategies in Emerging Architectures.

Academia and industry have proposed several architectures for multimedia streaming services over ICN. The dynamic adaptive streaming over popularity-driven caching (DASCache) [36] architecture is introduced to enable fast video streaming services with good QoE over ICNs where the average access time per bit of video content requested by the end-user is minimized. DASCache can also enable streaming clients to achieve the best video watching experience by switching to another video with better resolution during a video streaming session even in varying network conditions [36]. Ge et al. [43] propose a QoE-driven content caching and adaptation scheme over MEC-enabled ICN-based architecture where the information regarding video streaming clients and network capacity usage are used by the controller to prefetch video segments and move them the edge caches for users to download. The main goal of this approach is to enhance the users' QoE by managing and monitoring the popularities of both video

Table 7.1 A summary of QoE-centric management strategies in emerging architectures.

Strategy	Major contributions	Contribution/Objectives/Functionality
Application-level optimizations	Adaptive streaming over HTTP/2 [19–22], client-based prefetching [23], Meta-heuristics for increased client QoE-awareness [2, 24]	Improving the video quality and reduce the live latency using HTTP/2 protocol. HTTP/2 is a server push mechanism that also increase link utilization compared to HTTP/1.1 [20]. Meta-heuristics approaches exploit context information to improve the bitrate selection strategy of the client
QoE-aware/driven adaptive streaming over MEC	QoE-aware software driven multi-access edge service management [6, 8–10, 25–27]	Decrease the content delivery latency and improve the utilization of the network resources. Enable QoE monitoring and video quality adaptation using the real-time knowledge of UEs and network
QoE-aware adaptive streaming over cloud/fog computing	QoE-based resource management [28–33], QoE optimization with energy efficiency [34]	Perform QoS/QoE-aware orchestration of resources by scheduling flows between services. Some of the proposals such as in [35] can enable service providers to predict the QoE of DASH-supported video streaming using fog nodes
QoE-driven/aware management using ICN	[36, 37], QoE-driven content caching and adaptation scheme over MEC-enabled ICN [9]	Performing prefetching of video streams that enables ICN to compute the link resources availability and makes scheduling of data units dissemination called “chunks” to edge caches according to end-users’ requests [36] or video prefetching at the network edge in order to achieve the users’ QoE [4]

segments and their representations at the mobile network edge. A QoE- driven mobile edge caching placement optimization architecture is introduced in [9] to reduce total video distortion. The coordination among distributed edge servers and rate-distortion characteristics of multiple bitrate videos are considered in the implementation of the approach such that videos are accessed by users at a minimum latency. Yu et al. [37] propose a DASH-aware video stream prefetching architecture over ICN and Content Delivery Networks (CDNs). The network controller is applied to monitor network congestion, locate content, and manage the caches. A DASH-based architecture called Dynamic Adaptive Streaming over Content centric networking (DASC) that leverages the principles of CCN is proposed by Liu et al. [44] with the aim of overcoming the network bottlenecks by progressively downloading the contents to the best available video quality. The authors in [45] propose an NDN-based architecture that provides the interplay between different interest forwarding strategies and adaptation heuristics. The upper bound for the average video bitrate the clients can stream is achieved, assuming that the network states and video streaming characteristics of clients and are known a priori. Although ICN has shown potential for video streaming, the video rate adaptation of DASH clients can also be complicated because they are not aware of the node (e.g. the original server) that provides the video content. However, similar to the objectives of a regular DASH, intelligent caching strategies in ICN can be developed [37, 46] to optimize resource allocation from network cache nodes and enhance the video delivery performance for the end- users. It is important to mention that caching in ICN employs the naming structure of the video packet interests, a technique that enables the network to know in real-time he relationship between different video segments and what a DASH client is watching [1].

7.4 QoE-aware Adaptive Streaming over Cloud/Fog Computing

Fog computing [47, 48] is a distributed computing architectural approach that enables service providers to extend the cloud services to the network edge. Fog computing is an extension of cloud computing paradigm which also supports network virtualization functions, similar to MEC approach. However, different from cloudlet [49] and MEC paradigms [50], fog computing can not be implemented and operate as a standalone because it is highly linked to the existence of a cloud. Several SDN/NFV-architectures that can manage and orchestrate resources in multi-domain and multi-technology networks on top of a cloud/fog infrastructure have been proposed [29] and [30]. Rosário et al. [51] propose an SDN/NFV-based multi-tier fog computing strategy that enhances the delivery of video services to clients with QoE support through the cooperation between fog and cloud and fog. With this approach, the video contents/services are moved from cloud computing to fog nodes. That way, it allows mobile users to have access to various fog services with a low network controller overload and QoE support. Gupta et al. [28] introduce SDFog, a SDN-based fog computing architecture that orchestrate QoS/QoE-aware network resources by scheduling flows between services. SDFog is designed to enable an end-to-end resource management in a heterogeneous FC environment with QoS/QoE guarantees. The Distributed Service Orchestration Engine (DSOE) and Service Oriented Middleware (SOM) are the major components of SDFog. The DSOE is responsible for service discovery, flow creation and network parameter calculations, and QoS aware orchestration of resources by scheduling flows between the services [1, 28].

Zheng et al. [35] propose a Fog-assisted Real-time QoE Prediction (FRQP) architecture for service providers to perform QoE prediction for DASH-supported video streaming using fog computing nodes. A network probe at fog nodes is further designed and implemented to measure network traffic, collect packet header information and infer users' QoE according to the temporal features of the video traffic flows. MEdia FOg Resource Estimation (MeFoRE) is one of the QoE-based framework proposed by Aazam et al. [52] to provide resource estimation at Fog and to enhance QoS in IoT environments. Authors in [32] propose a joint optimization of energy that takes into account QoE -fairness mechanisms in cooperative fog computing. A QoE performance evaluation of Fog and cloud computing service orchestration architecture in future networks is introduced in [1] where throughput, latency, and energy efficiency are important parameters considered for QoE evaluation per user for different payloads. For a more recent survey of delay-sensitive video applications (e.g. video conferencing) in the cloud, we refer the reader to [1, 53].

7.5 Conclusion

The advancement of emerging architectures such as MEC, Cloud/Fog computing, and ICN have enabled service providers to move contents close to their customers and deliver services to them at a minimum latency. With cloud computing today, on-demand access to a shared pool of configurable computing resources (e.g. networks, servers, storage) with minimum management effort is possible. ICN can improve the end users' QoE through edge cache prefetching mechanisms that consider network monitoring for populating a cache as per the requirements/needs ahead of the client requests. The emergence of new services and applications in future networks such as 6G networks requires a 3D architecture to provide ubiquitous 3D coverage, seamless and extreme service connectivity, pervasive connectivity, unmanned mobility, and holographic telepresence support. To improve radio access capability and unlock the support of on-demand edge cloud services in 3D space, the future heterogeneous architecture design should complement and integrate with terrestrial networks and non-terrestrial platforms such as low-orbit satellites, balloons, and drones. The data-driven network and services management in future networks using AI/ML are other important aspects that bring value for service providers.

Bibliography

1 Barakabitze, A.A., Barman, N., Ahmad, A. et al. (2019). QoE management of multimedia services in future networks: a tutorial and survey. *IEEE Communication Surveys and Tutorials* 22 (1): 526–565.

2 Tselios, C. and Tsolis, G. (2016). On QoE-awareness through virtualized probes in 5G networks. *2016 IEEE 21st International Workshop on Computer Aided Modelling and Design of Communication Links and Networks (CAMAD)*, pp. 159–164, October 2016.

3 Li, S., Guo, Z., Shou, G. et al. (2016). QoE analysis of NFV-based mobile edge computing video application. *2016 IEEE International Conference on Network Infrastructure and Digital Content (IC-NIDC)*, pp. 411–415, September 2016. https://doi.org/10.1109/ICNIDC.2016.7974607.

4 Ge, C., Wang, N., Foster, G., and Wilson, M. (2017). Toward QoE-assured 4K video-on-demand delivery through mobile edge virtualization with adaptive prefetching. *IEEE Transactions on Multimedia* 19 (10): 2222–2237.

5 Ge, C. and Wang, N. (2018). Real-time QoE estimation of DASH-based mobile video applications through edge computing. *IEEE INFOCOM 2018 - IEEE Conference on Computer Communications Workshops (INFOCOM WKSHPS)*, pp. 766–771, Honolulu, HI, USA, April 2018.

6 Quadri, C., Gaito, S., Bruschi, R. et al. (2018). A MEC approach to improve QoE of video delivery service in urban spaces. *2018 IEEE International Conference on Smart Computing (SMARTCOMP)*, pp. 25–32, Taormina, Italy, June 2018. https://doi.org/10.1109/SMARTCOMP.2018.00095.

7 Li, Y., Frangoudis, P.A., Hadjadj-Aoul, Y., and Bertin, P. (2017). A mobile edge computing-assisted video delivery architecture for wireless heterogeneous networks. *2017 IEEE Symposium on Computers and Communications (ISCC)*, pp. 534–539, Heraklion, Greece, July 2017. https://doi.org/10.1109/ISCC.2017.8024583.

8 Ge, C., Wang, N., Chai, W.K., and Hellwagner, H. (2018). QoE-assured 4K HTTP live streaming via transient segment holding at mobile edge. *IEEE Journal on Selected Areas in Communications* 36 (8): 1816–1830.

9 Li, C., Toni, L., Zou, J. et al. (2018). QoE-driven mobile edge caching placement for adaptive video streaming. *IEEE Transactions on Multimedia* 20 (4): 965–984. https://doi.org/10.1109/TMM.2017.2757761.

10 Zhang, L., Wang, S., and Chang, R.N. (2018). QCSS: A QoE-aware control plane for adaptive streaming service over mobile edge computing infrastructures. *2018 IEEE International Conference on Web Services (ICWS)*, pp. 139–146, San Francisco, CA, USA, July 2018. https://doi.org/10.1109/ICWS.2018.00025.

11 Barakabitze, A.A., Ahmad, A., Mijumbi, R., and Hines, A. (2020). 5G network slicing using SDN and NFV: a survey of taxonomy, architectures and future challenges. *Computer Networks* 167: 1–40.

12 Agiwal, M., Roy, A., and Saxena, N. (2016). Next generation 5G wireless networks: a comprehensive survey. *IEEE Communication Surveys and Tutorials* 18 (3): 1617–1655.

13 Akyildiz, I.F., Kak, A., and Nie, S. (2020). 6G and beyond: the future of wireless communications systems. *IEEE Access* 8: 13399–134030.

14 Strinati E.C., Barbarossa, S., Choi, T. et al. (2020). 6G in the sky: on-demand intelligence at the edge of 3D networks (Invited paper). *Wiley ETRI Journal* 42 (5): 1–15.

15 Kim, J., Casati, G., Pietrabissa, A. et al. (2020). 5G-ALLSTAR: An integrated satellite-cellular system for 5G and beyond. *IEEE Wireless Communications and Networking Conference Workshops (WCNCW)*, pp. 1–6, October 2020.

16 Mendoza, F., Ferrus, R., and Sallent, O. (2014). Experimental proof of concept of an SDN-based traffic engineering solution for hybrid satellite-terrestrial mobile backhauling. *International Journal of Satellite Communications and Networking* 37 (3): 1–18.

17 Kim, J., Casati, G., Pietrabissa, A. et al. (2020). 5G-ALLSTAR: An integrated satellite-cellular system for 5G and beyond. *IEEE Wireless Communications and Networking Conference Workshops (WCNCW)*, pp. 1–6.

18 Shaat, M., Perez-Neira, A.I., Agapiou, G. et al. (2016). SANSA: Hybrid terrestrial-satellite backhaul network for the 5th generation. *ETSI Workshop on Future Radio Technologies – Air Interfaces*, pp. 1–6, January 2016.

19 Xiao, M., Swaminathan, V., Wei, S., and Chen, S. (2016). DASH2M: Exploring HTTP/2 for internet streaming to mobile devices. *Proceedings of the ACM on Multimedia Conference*, pp. 22–31.

20 Zhao, S. and Medhi, D. (2017). SDN-assisted adaptive streaming framework for tile-based immersive content using MPEG-DASH. *IEEE Conference on Network Function Virtualization and Software Defined Networks (NFV-SDN)*, November 2017.

21 van der Hooft, J., Petrangeli, S., Wauters, T. et al. (2017). An HTTP/2 push-based approach for low-latency live streaming with super-short segments. *Journal of Network and Systems Management* 26 (1): 51–78.

22 Nguyen, D.V., Le, H.T., Nam, P.N. et al. (2016). Request adaptation for adaptive streaming over HTTP/2. *IEEE International Conference on Consumer Electronics (ICCE)*, pp. 189–191, January 2016.

23 Krishnamoorthi, V., Carlsson, N., Eager, D. et al. (2015). Bandwidth-aware prefetching for proactive multi-video preloading and improved HAS performance. *23rd ACM International Conference on Multimedia (MM)*, pp. 551–560, May 2015.

24 van der Hooft, J., Petrangeli, S., Claeys, M. et al. (2015). A learning-based algorithm for improved bandwidth-awareness of adaptive streaming clients. *IFIP/IEEE International Symposium on Integrated Network Management (IM)*, pp. 131–138, May 2015.

25 Liang, C., He, Y., Yu, F.R., and Zhao, N. (2017). Enhancing QoE-aware wireless edge caching with software-defined wireless networks. *IEEE Transactions on Wireless Communications* 16 (10): 6912–6925.

26 Peng, S., Fajardo, J.O., Khodashenas, P.S. et al. (2017). QoE-oriented mobile edge service management leveraging SDN and NFV. *Mobile Information Systems* 2017 1–14.

27 Ravindran, R., Chakraborti, A., Amin, S.O. et al. (2017). 5G-ICN: Delivering ICN services over 5G using network slicing. *IEEE Communications Magazine* 55 (5): 101–107.

28 Gupta, H., Nath, S.B., Chakraborty, S., and Ghosh, S.K. (2017). SDFog: A software defined computing architecture for QoS aware service orchestration over edge devices, pp. 1–8, September 2017. *arXiv preprint; arXiv:1609.01190.*

29 Vilalta, R., Mayoral, A., Casellas, R. et al. (2016). SDN/NFV orchestration of multi-technology and multi-domain networks in Cloud/Fog architectures for 5G services. *21st OptoElectronics and Communications Conference (OECC) Held Jointly with 2016 International Conference on Photonics in Switching (PS)*, July 2016.

30 Vilalta, R., Mayoral, A., Casellas, R. et al. (2016). Experimental demonstration of distributed multi-tenant Cloud/Fog and heterogeneous SDN/NFV orchestration for 5G services. *European Conference on Networks and Communications (EuCNC)*, pp. 1–5, June 2016.

31 Chaudhary, R., Kumar, N., and Zeadally, S. (2017). Network service chaining in Fog and Cloud computing for the 5G environment: data management and security challenges. *IEEE Communications Magazine* 55 (11): 114–122.

32 Kitanov, S. and Janevski, T. (2016). Quality evaluation of Cloud and Fog orchestrated services in 5G mobile networks. *Enabling Technologies and Architectures for Next-Generation Networking Capabilities*, Volume 1, pp. 1–36, June 2016.

33 Mahmud, R., Srirama, S.N., Ramamohanarao, K., and Buyya, R. (2019). Quality of experience (QoE)-aware placement of applications in Fog computing environments. *Journal of Parallel and Distributed Computing* 132: 190–203. https://doi.org/https://doi.org/10.1016/j.jpdc.2018.03.004.

34 Dong, Y., Han, C., and Guo, S. (2018). Joint optimization of energy and QoE with fairness in cooperative Fog computing system. *2018 IEEE International Conference on Networking, Architecture and Storage (NAS)*, pp. 1–4, Chongqing, China, October 2018. https://doi.org/10.1109/NAS.2018.8515738.

35 Zheng, H., Zhao, Y., Lu, X., and Cao, R. (2016). A mobile Fog computing-assisted DASH QoE prediction scheme. *Wireless Communications and Mobile Computing* 2018: 1–10.

36 Li, W., Oteafy, S.M.A., and Hassanein, H.S. (2015). Dynamic adaptive streaming over popularity-driven caching in information-centric networks. *IEEE ICC 2015 - Next Generation Networking Symposium*, pp. 5747–5752, June 2015.

37 Yu, Y.-T., Bronzino, F., Fan, R. et al. (2015). Congestion-aware edge caching for adaptive video streaming in information-centric networks. *12th Annual IEEE Consumer Communications and Networking Conference (CCNC)*, 588–596, pp. January 2015.

38 Barakabitze, A.A. and Xiaoheng, T. (2014). Caching and data routing in information centric networking (ICN): the future internet perspective. *International Journal of Advanced Research in Computer Science and Software Engineering* 4 (11): 26–36.

39 Ahlgren, B., Dannewitz, C., Imbrenda, C. et al. (2012). A survey of information-centric networking. *IEEE Communications Magazine* 50 (7): 26–36.

40 Majeed, M.F., Ahmed, S.H., Muhammad, S. et al. (2017). Multimedia streaming in information-centric networking: a survey and future perspectives. *Computer Networks* 125: 103–121.

41 Zhang, L., Afanasyev, A., Burke, J. et al. (2014). Named data networking. *SIGCOMM Computer Communication Review* 44 (3): 66–73. https://doi.org/http://doi.acm.org/10.1145/2656877.2656887.

42 Koponen, T., Chawla, M., Chun, B. et al. (2007). A data-oriented (and beyond) network architecture. *ACM SIGCOMM Computer Communication Review* 37: 181–192.

43 Ge, C., Wang, N., Skillman, S. et al. (2016). QoE-driven DASH video caching and adaptation at 5G mobile edge. *Proceedings of the 3rd ACM Conference on Information-Centric Networking*, pp. 237–242, September 2016.

44 Liu, Y., Geurts, J., Point, J. et al. (2013). Dynamic adaptive streaming over CCN: a caching and overhead analysis. *2013 IEEE International Conference on Communications (ICC)*, pp. 3629–3633, June 2013. https://doi.org/10.1109/ICC.2013.6655116.

45 Rainer, B., Posch, D., and Hellwagner, H. (2016). Investigating the performance of pull-based dynamic adaptive streaming in NDN. *IEEE Journal on Selected Areas in Communications* 34 (8): 2130–2140.

46 Li, W., Oteafy, S., and Hassanein, H. (2017). Rate-selective caching for adaptive streaming over information-centric networks. *IEEE Transactions on Computers* 66 (9): 1613–1628.

47 Yi, S., Hao, Z., Qin, Z., and Li, Q. (2015). Fog computing: platform and applications. *Proceedings of the 3rd IEEE Workshop on Hot Topics in Web Systems and Technologies (HotWeb)*, pp. 73–78, November 2015.

48 Liu, Y., Fieldsend, J.E., and Min, G. (2017). A framework of fog computing: architecture, challenges, and optimization. *IEEE Access: Special Section on Cyber-Physical Social Computing and Networking* 5: 25445–25454.

49 Satyanarayanan, M., Bahl, P., Caceres, R., and Davies, N. (2009). The case for VM-based cloudlets in mobile computing. *IEEE Pervasive Computing* 8 (4): 14–23.

50 Pham, Q.-V., Fang, F., Vu, N.H. et al. (2020). A survey of multi-access edge computing in 5G and beyond: fundamentals, technology integration, and state-of-the-art. *IEEE Access* 8: 116974–117017.

51 Rosário, D., Schimuneck, M., Camargo, J. et al. (2018). Service migration from cloud to multi-tier Fog nodes for multimedia dissemination with QoE support. *Sensors* 18 (2): 1–17.

52 Aazam, M., St-Hilaire, M., Lung, C., and Lambadaris, I. (2016). MeFoRE: QoE based resource estimation at Fog to enhance QoS in IoT. *2016 23rd International Conference on Telecommunications (ICT)*, pp. 1–5, Thessaloniki, Greece, May 2016. https://doi.org/10.1109/ICT.2016.7500362.

53 Abdallah, M., Griwodz, C., Chen, K.-T. et al. (2018). Delay-sensitive video computing in the cloud: a survey. *ACM Transactions on Multimedia Computing, Communications, and Applications* 14 (3s): 54:1–54:29.

8

Emerging Applications and Services in Future 5G Networks

This chapter provides a comprehensive research study for QoE measurements in emerging applications focusing on immersive AV/VR, cloud video gaming, user Quality of Experience (QoE) in the context of the emerging multiple-sensorial media (mulsemedia) services, and 360° Immersive Video.

8.1 QoE in Immersive AR/VR and Mulsemedia Applications

The global market of 360° cameras grew at an annual rate of 35% from 2016 to 2020 while the global market for VR related devices reached 30 billion USD by 2020. With the increasing acceptance of and demand for 360° immersive videos which are of much higher size than traditional 2D videos, service providers will need to reduce End-to-End (E2E) network latency, bandwidth limits, and improve overall Quality of Service (QoS)/QoE for video streaming media services as new use cases for Augmented Reality (AR) and Virtual Reality (VR) are introduced to the market [1]. These issues not only affect consumers' AR/VR experiences today, but also pose a barrier to the delivery of more immersive real-time experiences in the future. Researchers from academia and industry have lately begun to quantify, model, and manage QoE when the user ingested content is beyond typical audio and video materials, in anticipation of the predicted rise in personal Head-Mounted Displays (HMDs) and other content generating devices. Viewport-dependent methods [2, 3] for 360° video streaming have been presented in these approaches because they can lower the bandwidth required to transmit the video. In VR, the user is immersed in a virtual environment and can choose their favorite viewpoint dynamically and freely [4]. In viewpoint-dependent streaming, the portion of the video beyond the view-point is delivered at a lower or average quality, while the piece of the video observed by the user is downloaded at the maximum quality possible.

Multimedia Streaming in SDN/NFV and 5G Networks: Machine Learning for Managing Big Data Streaming, First Edition. Alcardo Barakabitze and Andrew Hines.

Online transcoding approaches, such as foveate-based encoding [3] or spatially tiling the video [2], can be used to achieve viewpoint-dependent streaming. Because of their uses in diverse disciplines such as emergency response training and medical operations, QoE metrics have also been addressed in the context of AR systems [5]. Perritaz et al. [6] present a video bitrate adaption solution that adapts the frame rate and image size in real time to maximize the end users' QoE. Puig et al. [7] discuss QoE issues in AR applications, such as subjective quality assessment and psychophysics, as well as usability, ergonomics, human factors, and ethnography. Pallot et al. [8] suggest a taxonomy of QoE in augmented sports, as well as a 3D LIVE project that uses the QoS-UX-QoE strategy in three use scenarios (skiing, jogging, and golfing). Schatz et al. [9] investigate the influence of stalling occurrences in a fully immersive situation involving participants watching omnidirectional films using an HMD.

Academia and industry have been working on research regarding QoE in emerging mulsemedia services that involve various media objects compared to multimedia streaming services [10, 11]. Mulsemedia integrates human senses or 360° sensory media content, (for example olfactory, thermoceptics media objects and haptic) into human-computer interaction that eventually enable better QoE. The only drawbacks of 360° mulsemedia streaming services are the cost of more bandwidth than the conventional applications [12] and stringent delay requirements [13]. Yuan et al. introduce QoE of mulsemedia services where users can inform the mulsemedia server about both user preferences and network delivery conditions [14]. The authors in [15] provides a study regarding the impact of intensity of various mulsemedia components (e.g. air-flow and haptic) on the end user's perceived QoE. The results indicate that mulsemedia improves the user's overall QoE and enjoyment levels by up to 77%. Comsa et al. [16] present a novel 360° mulsemedia delivery system shown in Figure 8.1 where the 360° video capturing device can collect different olfactory types that are associated with

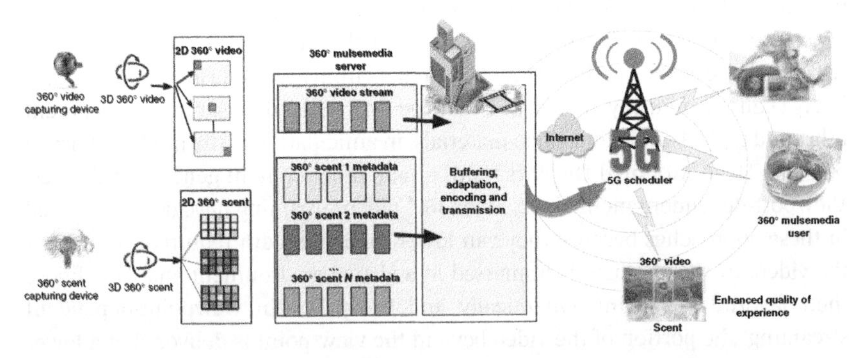

Figure 8.1 360° mulsemedia experience delivery system over next generation wireless networks.

the video representation. Several functionalists including the 360° media objects synchronization, 360° olfactory objects mapping, adaptation encoding, buffering and transmission are done by the 360° mulsemedia server. The 360° content is forwarded to the user's device through the 5G wireless networks where the radio scheduler's performance depends on mobility, positioning, channel conditions and the number of 360° mulsemedia users. Timmerer et al. [17] introduces the QoE aspects beyond audiovisual (sensory experience). The sensory effects (e.g. wind, vibration, ambient light) are used as an additional dimensions to contribute to the QoE of the end-users. The authors also provide a video dataset that is annotated with subjective user ratings and contains olfactory contents. A nonlinear model for QoE prediction of dynamic patio-temporal mulsemedia that consists of audiovisual sequences (with other sensory effects such as light, wind, vibration, and scent) is presented by Timmerer et al. [17].

360° VR video streaming content is gaining a lot of attention towards 6G networks in enabling MNO and content providers to attract customers at major events such as sports games, entertainment, etc. For example, using 360° live VR, users can participate in the entertainment events from anywhere without attending physically at the event station. QoE assessment regarding AR/VR has been extensively studied in the past [2, 3]. However, the VR-centric E2E network operation, monitoring, control and management architecture for future networks becomes important for understanding and enhancing the users' perceived QoE of immersive multimedia streaming. Therefore, the development of an intelligent QoE-centric E2E solutions based on multi-sensing techniques that can monitor, detect and demarcate faults in real-time is required. These solutions should model user experience and measure the perceivable media quality based on future softwarized network transmission capabilities [18].

8.2 QoE in Cloud Gaming Video Streaming Applications

With the advent of two connected and accessible gaming businesses toward 6G networks, gaming video streaming is growing in popularity. Passive gaming video streaming, for example, is a type of service in which players' gameplay is streamed from the client to streaming sites like YouTube-Gaming, Twitch.tv, and Facebook Gaming. The second type is a cloud gaming service, in which the game is rendered on the cloud and user gameplay is broadcast to the client in real time. Nvidia's GeforceNow is an example of a cloud gaming service [1]. Both gaming services have unique network requirements and limits, necessitating the use of various network management strategies. Cloud gaming is a multimedia application that is time-sensitive. The properties of passive gaming video streaming are similar to those of typical live/on-demand video streaming services. Cloud gaming may gain a lot from new emerging technologies like Software-Defined Networking (SDN) because it allows for better flow distribution.

Some academics have been working on improving the SDN controller's latency and, as a result, the quality of service. In order to optimize the flow distribution within the cloud gaming data center, Amiri et al. [19] proposed an optimization model by considering various factors such as the server load, game type and delay of the current path. The authors of [20] presented a Lagrangian Relaxation heuristic approach to decrease the complexity of the optimization method, which can then be applied in the data center using the OpenFlow controller [20]. Although SDN can improve delay-sensitive applications such as cloud gaming, more research into the various tools and strategies that can help reduce overall latency is needed. There are current standardization initiatives concerning the subjective and objective quality assessment of video gaming streaming services. The ITU-T recommendations regarding video gaming have been published in [21]. The standardization activities include (i) P.809 – that provides the definition of subjective methods to evaluate the QoE for cloud gaming services and (ii) G.1032 – for identifying different factors that affects QoE in gaming applications. ITU-T Study Group 12 established the G.OMG to develop a QoE-based gaming model that predicts the overall quality based on the E2E attributes of the user, system, network, and the context of video gaming usage. There are also a number of studies on the objective and subjective quality assessment of gaming video material, such as (i) evaluation of existing metrics [22–24], (ii) development of new no-reference metrics and models [25–27] for gaming content, and (iii) creation of gaming video datasets [28].

8.3 QoE in Light Field Applications

Light field videos, in comparison to traditional broadcast or streaming video content, typically have a very high resolution (about 50–80 megapixels) and hence a very large data need, necessitating more efficient compression as well as better network capacity [1]. In comparison with traditional stereoscopic displays, light field 3D displays necessitate distinct measurement processes since they demand the measurement of extra characteristics. This is due to the fact that geometrical optical techniques are used to create a hologram-like image. On multi-layer light field displays, Wang et al. [29] introduces QoE measurements for light field 3D displays. The authors also suggest three unique virtual models, namely the USAF-E model, the view angle model, and the concave/convex object model, to precisely evaluate spatial resolution, viewing angle, and depth resolution [1]. Shao et al. [30] propose a new full-reference quality assessment approach for stereoscopic pictures, in which binocular receptive field properties are aligned and learn with human visual perception. The authors in [31] investigated the perception of binocular quality in the setting of blurriness and blockiness. The luminance changes are taken into account while calculating the Global Luminance Similarity (GLS) index. The amplitude difference phase and amplitude difference of sparse coefficient vectors are used to generate the Sparse Feature Similarity (SFS) index [31]. When viewing rendered decompressed images, Perra [32] evaluates the QoE of light field applications. The authors further proposes a metric for evaluating the displayed views' quality that assesses the variation in structural similarity among a series of viewpoints derived from the light field [1].

8.4 Holographic and Future Media Communications

Holographic media [33] in 6G networks will need new form of communications over softwarized and virtualized systems such as holographic-type communications that are tolerant of quality degradation and characterized by very high throughput. Haptics and holograms will provide an immersive user experience even for multiple holographic streams. Hologram streaming in 6G networks will also support fast start-up and adapt to the changing network conditions in large bandwidth and low delay supported automated networks [34]. The 6G network will have new packetization models that support high precision for time-based and qualitative services to manage throughputs. Holographic and full-sensory immersive experiences will lead the applications in 6G networks and in a variety of market verticals. New network-friendly media formats will be characterized by mechanisms to disaggregate volumetric data sets to object centric approaches with lots of metadata support. New holographic applications are expected to emerge in 6G networks and provide fully immersive AR/VR/XR experience with holograms [35]. 6G networks will meet the requirements of extremely high data rates in the order of Gbps or Tbps and stimulate all human senses (vision, hearing, smell, taste, touch, and balance) that will be important for conveying real-time user experience. New distributed HTC techniques proposed in [36] that perform adaptive signaling and frame buffering can be a starting point towards designing efficient algorithms for managing and improving the user's QoE for teleportation streams.

8.5 Human-Centric Services and 3D Volumetric Video Streaming

Human-centered service will be another class of services that needs intelligent, trusted, and inclusive quality models by considering physical factors from human physiology (brain cognition, body physiology, and gestures) [35]. The future of human-centric service in 6G and beyond networks will be supported by ML to provide meaningful one-to-one streaming experiences for users and allow them to receive great services via communication channels over Human-centric Intelligentized Multimedia Networking (HIMN). ML will enable utilizing communication, computing, and caching resources at different edge devices of 6G multimedia networks. The human's experience in HIMN in 6G systems will be put at the center of mobile video streaming services. Optimization of video quality will be done in real-time based on the human's content characteristics, context-awareness and human's perceptions. New video codecs such as Versatile Video Coding (VVC) [37, 38] and other new video compression coding standards will play a central role in HIMN environments in 6G and beyond networks. Innovations in 5G and softwarized 6G networks provide new opportunities for human-centric video services for both consumers (e.g. cloud video gaming) and industries, particularly with respect to the multimedia sector. The video evolution toward 8K and beyond in 6G networks will rely on the strict requirements such as experienced data rate and low E2E latency for video delivery. However, with the development of new video coding standards (H.266)[1] for such services, a new set of Quality of Physical Experience (QoPE) metrics have to be defined and offered as mathematical function of traditional QoS and QoE metrics. The development of QoPE metrics models that learn human-brain can be achieved using AI/ML and multi-attribute utility theories from the operations research. A novel brain-aware learning and resource management approach proposed in [39] that explicitly factors in the brain state of human users during resource allocation in a cellular network can be a starting point in developing the QoPE models in 6G networks.

As we move toward 6G and beyond networks, multimedia content is not only gaining higher video resolutions but also higher degrees of immersion. Volumetric video streaming is an emerging key technology to offer user interactions and immersive representation of 3D spaces and objects in 6G and beyond networks [40]. Volumetric videos provide viewers a six Degree-of-Freedom (6DoF) and 3D rendering, making them highly immersive, interactive, and expressive. 6DoF means: three rotational dimensions (e.g. viewing direction in yaw, pitch, and roll) and three translational dimensions (e.g. viewpoint position in X, Y, and Z). Volumetric videos allow a viewer to freely change both the position in space and the orientation [41]. Streaming volumetric videos is highly bandwidth-demanding and requires lots of computational power because of their truly immersive nature. Thanks to the experienced data rate and peak data rate (≥1 Tb/s) offered by 6G networks which will play a significant part in delivering enough data fast enough for applications like volumetric video. Zhang et al. [42] propose a viewport prediction and blockage mitigation approaches that are efficient for

1 https://jvet.hhi.fraunhofer.de/.

streaming high-quality volumetric videos to multiple users. The authors provide a viewport-similarity opportunity that the multimedia research community can employ for optimizing effectively the network resource utilization using efficient multicast, and mmWave-aware multi-user video rate adaptation. Qian et al. [41] introduce an efficient resource volumetric video streaming architecture that leverages edge computing over 5G/6G to reduce the computation burden on smartphone users while maintaining a high QoE. While the QoE metrics for regular videos have been well studied, this remains an open problem for volumetric video streaming because the QoE of volumetric video streaming can be affected by factors such as viewing distance, visibility, point density, artifacts incurred by patches, motion-to-photon delay, and point density.

8.6 New Video Compression Standards Toward 6G Networks

The emerging use cases in 5G and 6G networks will need new video compression standards that offer enhanced capabilities compared to the video codecs in use today (H.264/MPEG-4 AVC, Google VP9, MPEG Dynamic Adaptive Streaming over HTTP (MPEG-DASH and others) [43]. The new video compression-coding standard, the VVC [38] has been recently standardized in a joint effort by the Moving Picture Experts Group of the ITU-T and MPEG of the ISO/IEC. VVC has been registered as ITU-T Recommendation H.266 — ISO/IEC 23090-3 to provide new pathways and a suitable performance level for new multimedia services over softwarized 5Gand beyond networks. New features of H.266 include (i) the highest compression efficiency compared to other video codecs, and new features to enhance support for low-delay video coding and immersive video, (ii) VVC achieves a reduction in bitrate of around 40% for existing HD and 4K/8K video services deployed with High Efficiency Video Coding (HEVC), at the same visual quality, and (iii) VVC is suitable for higher resolution video streams that can operate with 64×64 sample size transforms and block sizes of up to 128×128 pixels [44]. VVC will be the key for immersive video and time-critical video applications (e.g. videoconferencing, cloud video gaming, and remote control of road vehicles and drones), which will be prevalent in 6G and beyond networks. Figure 8.2 shows different application areas of VVC in 6G and beyond networks such as immersive XR and telepresence and extremely low latency cloud gaming, etc. It is worth mentioning that VVC offers specialized tools for the coding of the screen content and the computer-generated content of applications (e.g. cloud-based collaboration, cloud gaming, remote screen sharing). Furthermore, VVC offers an independent sub-pictures for applications such as tiled streaming of 360° video, enabling higher resolution for the portion of

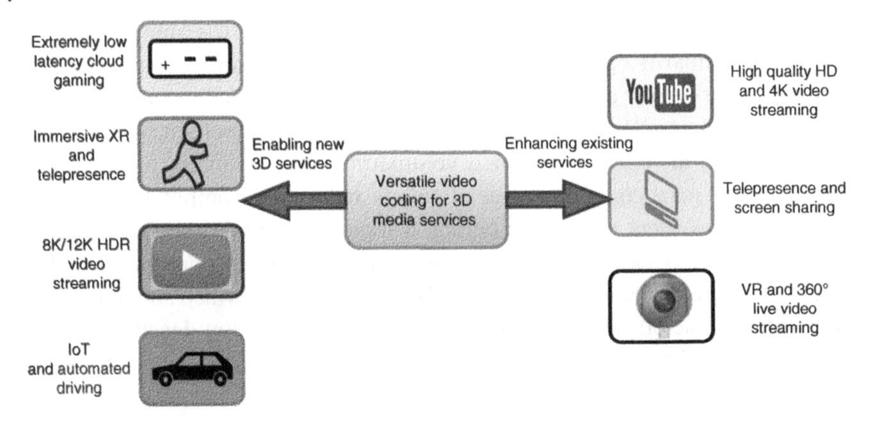

Figure 8.2 Different application areas of VVC in 6G networks.

a 360° video in view. Grois et al. [45] provide a performance comparison of H.265, VVC and MPEG-5 Essential Video Coding (EVC) in terms of computational complexity and coding gains. The results indicate that EVC provides 30% bitrate savings compared to HEVC for encoding 4K/2160p entertainment content (such as Video on Demand [VoD]) while VVC provides larger bitrate savings of about 40%. VVC is a promising successor to HEVC in terms of coding gains though at a price of significant computational complexity. HoangVan et al. [46] propose a VVC-based quality scalability strategy which offers higher compression efficiency using a layer referencing approach where the base and enhancement layers decoded information are jointly exploited to create a new enhancement layers coding reference. Mansri et al. propose a performance evaluation comparison of H.264, AVI, VP9, VVC and HEVC encoders for low-delay video services and applications. While the VP9 provides average bit-rate overhead, VVC offers the highest compression performance compared to the most relevant benchmarks in video coding industrial standards. The VVC standard advances the state of the art of video compression and will mark the conclusion of the last decade of video coding standardization because of its unprecedented application versatility.

8.7 Conclusion

VR and AR have been recognized and approved by the industry and academia as future applications that will provide end-users with genuinely immersive and interactive multimedia experiences. AR HMDs with mobile compatibility (e.g. the Epson Moverio BT-300, Samsung Gear VR, and Microsoft HoloLens), can allow users to experience their real-world environments. This chapter indicates that the multimedia community has shown a strong desire to investigate QoE aspects in new areas such as light field applications, immersive mulsemedia, AR/VR and video gaming. Video gaming experiences will soon evolve beyond our imaginations with the unprecedented power and capabilities that 5G brings. The majority of the methodologies presented in this chapter determines the viewers' QoE for watching videos on an AR/VR device. In AR applications, some of the works look at psychophysics and subjective quality assessment as well as human factors, ergonomics, usability, ethnography, and other topics. This chapter also highlighted the growing notion of 360° mulsemedia as an application that will transform streaming technology in future 5G networks. The chapter has also presented the assessment of QoE for light field applications that use 3D displays on multilayer LCDs or binocular receptive field qualities. The majority of works towards new domain of 6G applications and services has been in investigating the haptics and holograms streaming, 3D volumetric streaming, human-centric services and new video compression standards such as the Video Coding (VVC) that has been recently standardized in a joint effort by the Moving Picture Experts Group of the ITU-T and MPEG of the ISO/IEC.

Bibliography

1 Barakabitze, A.A., Barman, N., Ahmad, A. et al. (2019). QoE management of multimedia services in future networks: a tutorial and survey. *IEEE Communication Surveys and Tutorials.*

2 Qian, F., Ji, L., Han, B., and Gopalakrishnan, V. (2016). Optimizing 360 video delivery over cellular networks. *Proceedings of the 5th Workshop on All Things Cellular: Operations, Applications and Challenges (ATC'16)*, pp. 1–6, New York, NY, May 2016.

3 Ryoo, J., Yun, K., Samaras, D. et al. (2016). Design and evaluation of a foveated video streaming service for commodity client devices. *Proceedings of the 7th International Conference on Multimedia Systems (MMSys'16)*, pp. 1–6, Klagenfurt, Austria, May 2016.

4 Petrangeli, S., Van Der Hooft, J., Wauters, T., and De Turck, F. (2018). Quality of experience-centric management of adaptive video streaming services: status and challenges. *ACM Transactions on Multimedia Computing, Communications, and Applications (TOMM)* 14 (2s): 2–28.

5 Keighrey, C., Flynn, R., Murray, S., and Murray, N. (2020). A physiology-based QoE comparison of interactive augmented reality, virtual reality and tablet-based applications. *IEEE Transactions on Multimedia* 23: 333–341.

6 Perritaz, D., Salzmann, C., and Gillet, D. (2009). Quality of experience for adaptation in augmented reality. *IEEE International Conference on Systems, Man and Cybernetics*, pp. 888–893, December 2009.
7 Puig, J., Perkis, A., Lindseth, F., and Ebrahimi, T. (2012). Towards an efficient methodology for evaluation of quality of experience in augmented reality. *2012 4th International Workshop on Quality of Multimedia Experience*, pp. 188–193, July 2012. https://doi.org/10.1109/QoMEX.2012.6263864.
8 Pallot, M., Eynard, R., Poussard, B. et al. (2013). Augmented sport: exploring collective user experience. *Proceedings of the Virtual Reality International Conference (VRIC'13)*, pp. 1–10, Laval, France.
9 Schatz, R., Sackl, A., Timmerer, C., and Gardlo, B. (2017). Towards subjective quality of experience assessment for omnidirectional video streaming. *2017 9th International Conference on Quality of Multimedia Experience (QoMEX)*, pp. 1–6, Erfurt, Germany, May 2017. https://doi.org/10.1109/QoMEX.2017.7965657.
10 Covaci, A., Zou, L., Tal, I. et al. (2018). Is multimedia multisensorial? - A review of mulsemedia systems. *ACM Computing Surveys (CSUR)* 51 (5). pp. 1–35
11 Sulema, Y. (2016). Mulsemedia vs. multimedia: state of the art and future trends. *2016 International Conference on Systems, Signals and Image Processing (IWSSIP)*, pp. 1–5, May 2016. https://doi.org/10.1109/IWSSIP.2016.7502696.
12 Liu, T.-M., Ju, C.-C., Huang, Y.-H. et al. (2017). A 360-degree 4K×2K panoramic video processing Over Smart-phones. *2017 IEEE International Conference on Consumer Electronics (ICCE)*, pp. 247–249, January 2017. https://doi.org/10.1109/ICCE.2017.7889303.
13 Corbillon, X., Simon, G., Devlic, A., and Chakareski, J. (2017). Viewport-adaptive navigable 360-degree video delivery. *2017 IEEE International Conference on Communications (ICC)*, pp. 1–7, Paris, France, May 2017. https://doi.org/10.1109/ICC.2017.7996611.
14 Yuan, Z., Ghinea, G., and Muntean, G. (2014). Quality of experience study for multiple sensorial media delivery. *2014 International Wireless Communications and Mobile Computing Conference (IWCMC)*, pp. 1142–1146, August 2014. https://doi.org/10.1109/IWCMC.2014.6906515.
15 Yuan, Z., Chen, S., Ghinea, G., and Muntean, G.-M. (2014). User quality of experience of mulsemedia applications. *ACM Transactions on Multimedia Computing, Communications, and Applications* 11 (1s): 15:1–15:19. http://doi.acm.org/10.1145/2661329.
16 Comsa, I.-S., Trestian, R., and Ghinea, G. (2018). 360-degrees mulsemedia experience over next generation wireless networks - a reinforcement learning approach. *10th International Conference on Quality of Multimedia Experience (QoMEX 2018)*.

17 Timmerer, C., Waltl, M., Rainer, B., and Murray, N. (2014). *Sensory Experience: Quality of Experience Beyond Audio-Visual*, pp. 351–365. Cham: Springer International Publishing. ISBN 978-3-319-02681-7. https://doi.org/10.1007/978-3-319-02681-7_24.

18 Huawei Technologies: Virtual Reality/Augmented Reality White Paper. https://www-file.huawei.com/-/media/CORPORATE/PDF/ilab/vr-ar-en.pdf/ (Online: Accessed 13 June 2019).

19 Amiri, M., Al Osman, H., Shirmohammadi, S., and Abdallah, M. (2015). SDN-based game-aware network management for cloud gaming. *International Workshop on Network and Systems Support for Games (NetGames)*, p. 3.

20 Amiri, M., Al Osman, H., Shirmohammadi, S., and Abdallah, M. (2016). Toward delay-efficient game-aware data centers for cloud gaming. *ACM Transactions on Multimedia Computing, Communications, and Applications (TOMM)* 12 (5s): 71.

21 Möller, S., Schmidt, S., and Zadtootaghaj, S. (2018). New ITU-T standards for gaming QoE evaluation and management. *2018 10th International Conference on Quality of Multimedia Experience (QoMEX)*, pp. 1–6, May 2018. https://doi.org/10.1109/QoMEX.2018.8463404.

22 Barman, N. and Martini, M.G. (2021). User generated HDR gaming video streaming: dataset, codec comparison and challenges. *IEEE Transactions on Circuits and Systems for Video Technology* 32 (3): 1236–1249.

23 Zadtootaghaj, S., Schmidt, S., and Möller, S.. Modeling gaming QoE: towards the impact of frame rate and bit rate on cloud gaming. *2018 10th International Conference on Quality of Multimedia Experience (QoMEX)*, pp. 1–6. IEEE, 2018.

24 Barman, N., Zadtootaghaj, S., Martini, M.G. et al. (2018). A comparative quality assessment study for gaming and non-gaming videos. *2018 10th International Conference on Quality of Multimedia Experience (QoMEX)*, pp. 1–6, May 2018. https://doi.org/10.1109/QoMEX.2018.8463403.

25 Barman, N., Jammeh, E., Ghorashi, S.A., and Martini, M.G. (2019). No-reference video quality estimation based on machine learning for passive gaming video streaming applications. *IEEE Access* 7: 74511–74527. https://doi.org/10.1109/ACCESS.2019.2920477.

26 Utke, M., Zadtootaghaj, S., Schmidt, S., and Bosse, S. (2020). NDNetGaming - Development of a no-reference deep CNN for gaming video quality prediction. *Multimedia Tools and Applications*. https://doi.org/10.1007/s11042-020-09144-6.

27 Göring, S., Rao, R.R.R., and Raake, A. (2019). nofu — A Lightweight No-Reference Pixel Based Video Quality Model for Gaming Content. *2019 11th International Conference on Quality of Multimedia Experience (QoMEX)*, pp. 1–6, June 2019. https://doi.org/10.1109/QoMEX.2019.8743262.

28 Barman, N., Zadtootaghaj, S., Schmidt, S. et al. (2018). GamingVideoSET: A dataset for gaming video streaming applications. *2018 16th Annual Workshop* *on Network and Systems Support for Games (NetGames)*, pp. 1–6, June 2018. https://doi.org/10.1109/NetGames.2018.8463362.

29 Wang, S., Ong, K.S., Surman, P. et al. (2016). Quality of experience measurement for light field 3D displays on multilayer LCDs. *Society for Information Display* 24 (12): 726–740.

30 Shao, F., Li, K., Lin, W. et al. (2015). Full-reference quality assessment of stereoscopic images by learning binocular receptive field properties. *IEEE Transactions on Image Processing* 24 (10): 2971–2983.

31 Chen, M.-J., Cormack, L.K., and Bovik, A.C. (2013). No-reference quality assessment of natural stereopairs. *IEEE Transactions on Image Processing* 22 (9): 3379–3391.

32 Perra, C. (2018). Assessing the quality of experience in viewing rendered decompressed light fields. *Multimedia Tools and Applications* 77 (16): 21771–21790.

33 Clemm, A., Vega, M.T., Ravuri, H.K., T., Wauters (2015). Toward truly immersive holographic-type communication: challenges and solutions. *IEEE Communications Magazine* 58 (1): 93–99.

34 Network 2030: A Blueprint of Technology, Applications and Market Drivers Towards the Year 2030 and Beyond. https://www.itu.int/en/ITU-T/focusgroups/net2030/Documents/White_Paper.pdf (accessed 3 May 2022).

35 Letaief, K.B., Chen, W., Shi, Y. et al. (2019). The roadmap to 6G: AI empowered wireless networks. *IEEE Communications Magazine* 57 (8): 84–90.

36 Selinis, I., Wang, N., Da, B. et al. (2020). On the internet-scale streaming of holographic-type content with assured user quality of experiences. *IFIP Networking Conference (Networking)*, July 2020.

37 New ‘Versatile Video Coding’ standard to enable next-generation video compression. https://www.itu.int/en/mediacentre/Pages/pr13-2020-New-Versatile-Video-coding-standard-video-compression.aspx (accessed 3 May 2022).

38 Huang, Y.-W., Hsu, C.-W., Che, C.-Y. et al. (2020). A VVC proposal with quaternary tree plus binary-ternary tree coding block structure and advanced coding techniques. *IEEE Transactions on Circuits and Systems for Video Technology* 30 (5): 1311–1325.

39 Kasgari, A.T.Z., Saad, W., and Debbah, M. (2019). Human-in-the-loop wireless communications: machine learning and brain-aware resource management. *IEEE Transactions on Communications* 43 (11): 7727–7743.

40 Serhan, G., Podborski, D., Buchholz, T. et al. (2020). Low-latency cloud-based volumetric video streaming using head motion prediction. *Proceedings of the 30th ACM Workshop on Network and Operating Systems Support for Digital Audio and Video*, pp. 27–33, June 2020.

41 Qian, F., Han, B., Pair, J., and Gopalakrishnan, V. (2019). Toward practical volumetric video streaming on commodity smartphones. *Proceedings of the*

20th International Workshop on Mobile Computing Systems and Applications, pp. 135–140, February 2019.

42 Zhang, A., Zhang, A., Wang, C., Liu, X., B., Han (2021). Mobile volumetric video streaming enhanced by super resolution. *Proceedings of the 20th ACM Workshop on Hot Topics in Networks*, pp. 16–22, November 2021.

43 Barakabitze, A.A., Barman, N., Ahmad, A. et al. (2019). QoE management of multimedia streaming services in future networks: a tutorial and survey. *IEEE Communication Surveys and Tutorials* 22 (1): 526–565.

44 Sjoberg, R., Strom, J., ?. Litwic, and Andersson, K. (2020). Versatile video coding explained – the future of video in a 5G world. https://www.ericsson.com/4a92d7/assets/local/reports-papers/ericsson-technology-review/docs/2020/versatile-video-coding-explained.pdf (accessed 3 May 2022).

45 Grois, D., Giladi, A., Choi, K., M.W., park (2021). Performance comparison of emerging EVC and VVC video coding standards with HEVC and AV1. *SMPTE Motion Imaging Journal* 130 (4): 1–12.

46 HoangVan, X., NguyenQuang, S., F., Pereira (2020). Versatile video coding based quality scalability with joint layer reference. *IEEE Signal Processing Letters* 27: 2079–2083.

9

5G Network Slicing Management Architectures and Implementations for Multimedia

This chapter presents the concept of 5G network slicing in the contexts of multimedia applications. It provides a comprehensive description of multimedia-oriented sharing/slicing and service customization mechanisms in 5G networks. It also provides 5G network slicing proof of concepts and research projects in terms of their architectures and different implementation details. This chapter presents the management and orchestration of network slices in a single domain and multiple domains while supporting multi-tenants and multi-operators. Moreover, it presents comprehensive details of orchestrators which are becoming complementary to allow fast innovation in 5G using Software-Defined Network (SDN) and Network Function Virtualization (NFV) to realize the 5G slicing network concept.

9.1 5G Network Slicing Architectures and Implementations

From an industrial and research standpoint, 5G has been designed as a network that can meet the needs of various vectors while also meeting the service quality preferences of end-users. Projects such as SELFNET [1] and 5G-Xhaul [2] have been actively focusing on 5G and beyond network capabilities such as self-configuration, self-healing, and self-optimization. As the 5G network slicing concept matures, several standards bodies and associations (e.g. ITU Telecommunication [ITU-T], European Telecommunications Standards Institute [ETSI], Next Generation Mobile Networks [NGMN], 3rd Generation Partnership Project [3GPP]), academic and industrial research programs (5G Novel Radio Multiservice adaptive network Architecture [5G-NORMA], 5GEx), and vendors are all working on different goals in tandem, with some of them collaborating closely with the ETSI. This part provides the research projects regarding 5G network slicing in terms of their implementation architectures.

Multimedia Streaming in SDN/NFV and 5G Networks: Machine Learning for Managing Big Data Streaming, First Edition. Alcardo Barakabitze and Andrew Hines.

9.1.1 Collaborative 5G Network Slicing Research Projects

The MATILDA project [3] designs and implements a holistic 5G End-to-End (E2E) services operational framework tackling the lifecycle of design, development and orchestration of 5G-ready applications and 5G network services over programmable infrastructure, following a unified programmability model and a set of control abstractions. The project devises and realizes a radical shift in the development of software for 5G-ready applications as well as virtual and physical network functions and network services. This is achieved by defining proper abstractions, adopting a unified programmability model, and the creation of an open development environment that may be used by application and network functions developers [4]. SliceNet [5] designs an E2E cognitive network slicing and slice management framework in virtualized multi-domain, multi-tenant 5G networks. SliceNet aims toward the maximization of network resources sharing within and across different administrative domains. SliceNet presents an integrated Fault, Configuration, Accounting, Performance, Security (FCAPS) framework for truly E2E management, control and orchestration of slices by secured, interoperable, and reliable operations across multi-operator domains. 5G Exchange (5GEx) [6] project enables operators to collaborate on 5G infrastructure services in order to introduce unification via NFV/SDN compatible multi-domain orchestration by developing an open platform that enables cross-domain orchestration of services across various domains.

The basic element of the 5GEx infrastructure is a slice that uses lower-level 5GEx basic services and SDN/NFV approaches to efficiently serve 5G verticals. As demonstrated in Figure 9.1, standard interfaces are utilized to connect and communicate information among entities. The customers' 5GEx service requests are translated into a chain of Virtual Network Functions (VNFs) with their associated resource requirements using the Multi-domain Orchestrator (MdO) interface (1). Between 5GEx-enabled orchestrators, interface 2 trade slices in line with ELAs/ELAs and 5GEx higher-level services. Through interface 2, interface 3 is in charge of managing own or leased resources [7]. As shown in Figure 9.1, the customer-facing "third party orchestrator" is a virtual mobile network operator that implements MdO capability but does not own infrastructure. The 5Gex platform offers a number of collaborative modes, including "Direct peering" enabling multi-party remote cooperation. It also allows higher-level abstractions and complex models that span many exchange points or Points of Presence (PoPs) and cover views, resources, and services.

SONATA project [8] streamlines the development with abstract programming models, Service Development Kit (SDK) and a DevOps model that integrates operators, manufacturers, and third-party developers. SONATA supports the full service lifecycle including: development, testing, orchestration, deployment,

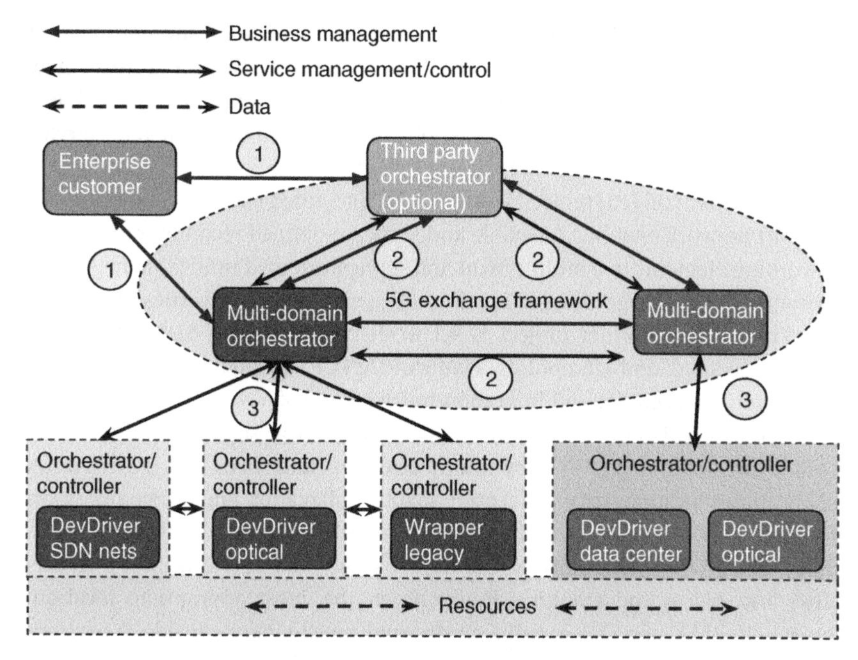

Figure 9.1 5GEx network slicing conceptual architecture.

management, and operations, and will define a roadmap for uptake of its results toward stakeholders' larger transition to SDN/NFV. 5G-MoNArch uses network slicing approach to design and implement a flexible, adaptable, and programmable 5G architecture that can support a wide range of use cases in vertical industries (e.g. healthcare, automotive, and media). The core component of the architecture of 5G-MoNArch is the M & O layer that employs the management and orchestration of network resources using the 3GPP specifications [9, 10]. Through a lightweight control plane and data plane programmability, the 5G!PAGODA project architecture provides effective network slice orchestration and management techniques in dispersed, edge-dominated network infrastructures. The project develops a scalable 5G slicing architecture that extends the current NFV architecture toward supporting different specialized network slices composed of multi-vendor VNFs, taking into account interoperability within/among the network slices and with legacy system-based services. The NECOS [11] project applies the Cloud Slicing (CS)[1] concept to propose the Lightweight Slice Defined Cloud (LSDC), an approach that extends the virtualization to all the resources

1 CS provides concurrent deployment of multiple logical, self-contained, independent, shared or partitioned slices on a common infrastructure platform. CS also enables dynamic multi-service and multi-tenancy support for 5G vertical market players.

in the involved networks and data centers and provides a uniform management with a high-level of orchestration. The NACOS project also introduces a new deployment models, the Slice as a Service (SaaS) where a slice is regarded as a grouping of resources managed as a whole, and that can accommodate service components, independent of other slices (Table 9.1).

The 5G-Crosshaul [15] project develops a 5G integrated backhaul and fronthaul transport network enabling a flexible and software-defined reconfiguration of all networking elements in a multi-tenant and service-oriented unified management environment to manage slices tailored to the specific needs of vertical industries [16]. The 5G-Transformer project is set to deliver a scalable SDN/NFV-based Mobile Transport and Computing Platform (MTCP) by adding the support of (i) a dynamic placement and migration mechanisms of VNFs, (ii) an integrated MEC services, (iii) new abstraction models for vertical services, and (iv) new mechanisms for sharing VNFs by multiple tenants and slices [17]. 5G-NORMA [12] proposes an innovative concept of adaptive allocation and (de)composition of mobile network functions using a Software-Defined Mobile Network Orchestration (SDMO), which flexibly decomposes the mobile Network Functions (NFs) and places the resulting functions in the most appropriate location. The 5G-NORMA provides a multi-service and multi-tenant capable 5G system architecture that enhances network programmability through a Software-Defined Mobile Network Control (SDMC) to efficiently handle the diverse requirements and traffic demand fluctuations. The fundamental entities (Edge Cloud, Network Cloud, and Controller) of the 5G-NORMA architecture is indicated in Figure 9.2. The *Edge Cloud* consists of the remote controllers and bases stations deployed at the radio or aggregation network, whereas the *Network Cloud* indicates one or more data-centers deployed at central sites. The *Controller* is responsible for executing and organizing the network functions that are co-located in the network cloud. The three innovative functionalities and five main pillars (A–E) of the 5G NORMA architecture are illustrated in Figure 9.2. Pillar A provides dynamic allocation and adaptive decomposition of mobile NFs between the network and edge cloud while pillar B signifies the SDMO that support multi-service and multi-tenancy and orchestrates network resources between different slices that belongs to different administrative domains. Pillar C indicates the joint optimization of both mobile access and core NFs localized together, either in the edge cloud or the network cloud. Pillar D includes the two innovative aspects of 5G NORMA functionalities. This include the handling of a variety of services and their accompanying QoS/QoE needs, as well as allows multi-service and context-aware customization of NFs. The mobile network multi-tenancy is highlighted in pillar E, which allows for the on-demand deployment of radio and core resources to virtual operators and vertical market players [7]. The 5G NORMA provides flexible connectivity of 5G networks using six building blocks

Table 9.1 A summary of academia/industry 5G projects and implementation based on SDN/NFV.

Name	Focus area			SDN/NFV related work
	SDN	NFV	QoE	
5G-NORMA [12]	Yes	Yes	Yes	Multi-service and context-aware adaptation of network functions to support a variety of services and corresponding QoE/QoS requirements
5G-MEDIA	Yes	Yes	Yes	A flexible network architecture that provides dynamic and flexible UHD (4K/8K) content distribution over 5G CDNs
5G-MoNArch	Yes	Yes	Yes	Employ network slicing to support the orchestration of both access and core network functions, and analytics, to support a variety of use cases in vertical industries such as automotive, healthcare, and media
5GTANGO	Yes	Yes	Yes	To develop a flexible 5G programmable network with an NFV-enabled Service Development Kit (SDK) that supports the creation and composition of VNFs and application elements as *"Network Services"*
SESSAME	Yes	Yes	Yes	Develop programmable 5G network infrastructure that support multi-tenancy, decrease network management OPEX while increasing the QoS/QoE and security
MATILDA	Yes	Yes	No	Orchestration of 5G-ready applications and network services over sliced programmable platforms
5G-Transformer	Yes	Yes	No	Develop an SDN/NFV-based 5G network architecture that meet specific vertical industries' (e.g. eHealth, automotive, industry 4.0, and media) requirements
5G-Crosshaul [13]	Yes	Yes	Yes	The design of 5G transport architectural solution that supports multi-domain orchestration among multiple network operators or service providers (e.g. multiple tenants)
5G-XHaul	Yes	Yes	Yes	Develop a scalable SDN control plane and mobility aware demand prediction models for optical/wireless 5G networks
CogNET [14]	Yes	Yes	No	Dynamic adaptation of network resources of VNFs, while minimizing performance degradations to fulfill SLA/ELAs requirements

(*Continued*)

Table 9.1 (Continued)

Name	Focus area			SDN/NFV related work
	SDN	NFV	QoE	
CHARISMA	Yes	Yes	Yes	Develop a software-defined converged fixed 5G mobile network architecture that offers both, multi-technology and multi-operator features
SaT5G	Yes	Yes	Yes	Integrated management and orchestration of network slices in 5G SDN/NFV-based satellite networks
SLICENET	Yes	Yes	Yes	Develop a cognitive network control, management and orchestration framework, that supports infrastructure sharing across multiple operator domains in SDN/NFV-enabled 5G networks
SONATA	Yes	Yes	Yes	Enable an integrated management and control to be part of the dynamic design of the softwarized 5G network architecture
COHERENT	Yes	Yes	No	Efficient radio resource modelling and management in programmable radio access networks
5G Exchange [6]	Yes	Yes	No	Enabling cross-domain orchestration of services over multiple administrations or over multi-domain single administrations

as extensively described in [7]: the Orchestrator (SDM-O), Coordinator (SDM-X), Software-Defined Mobile Network Controller (SDM-C), Mobility Management module, the QoE/QoS Mapping and Monitoring module.

9.1.2 Open Source Orchestrators, Proof of Concepts, and Standardization Efforts

To allow for rapid innovation, orchestrators for 5G network slicing are becoming more common, which is why the majority of the current solutions are open-source. An Orchestrator is software that can create, monitor and deploy network resources and services in the underlying softwarized environment. There are two types of orchestrators [18] as defined by the European Telecommunication Standard Institute (ETSI): (i) Resource Orchestrator (RO) and the (ii) Service Orchestrator (SO). The SO is responsible for creating E2E service between VNFs while the RO coordinates, authorizes, releases and engages Network Function Virtualization Infrastructure (NFVI) resources among different PoPs or within

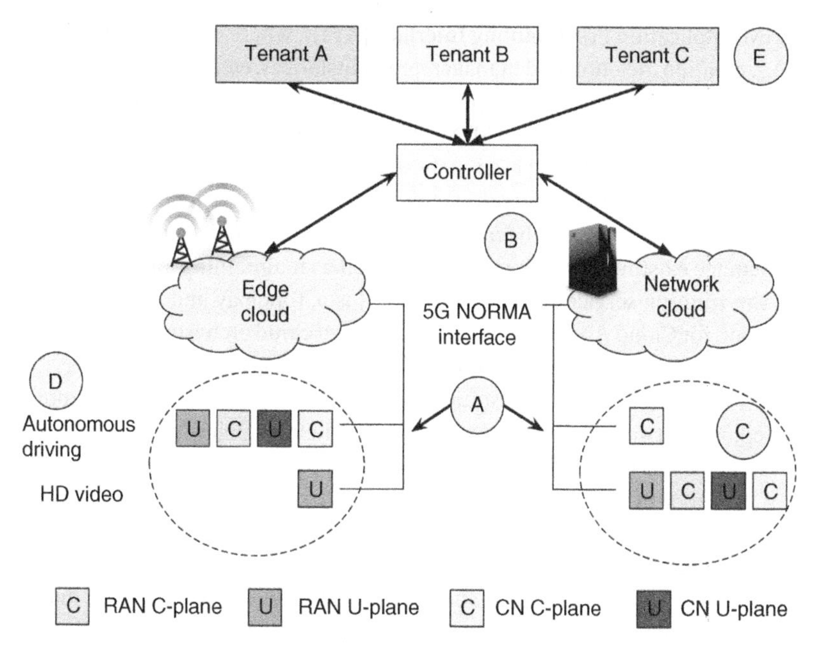

Figure 9.2 The main innovations of 5G NORMA concept.

one PoP. To realize the dynamic network slices orchestration and management in 5G and beyond networks, various orchestrators have been developed from academia and industry including [19]: Open Source NFV Management and Orchestration (MANO) (OSM), OpenMANO, OpenNFV, CloudNFV, OpenBaton, Cloudify, T-NOVA, Zero-time Orchestration, Operations and Management (ZOOM), OPNFV, ExperiaSphere, M-CORD, NGSON, and ONAP.

OSM is an open source MANO stack that is aligned with the ETSI NFV information models. OSM delivers a production-quality MANO stack that meets operators' requirements for commercial NFV deployments. The key aspect of OSM is the ability to orchestrate NFs of different natures: physical, virtual, and containerized. The OSM Release ELEVEN brings a set of new features including (i) support of standardized VNF and NS package formats, (ii) better integration across Physical Network Function (PNF), VNF, and Containerized Network Functions (CNFs), (iii) additional public cloud support, and (iv) CNF monitoring from Kubernetes metrics. OpenMANO is an open source project that provides a practical implementation of the MANO reference architecture that is currently being standardized by ETSI's NFV ISG (NFV MANO) [7]. It consists of three main components, namely: openMANO (a reference implementation of an NFV-O), openVIM, and openMANO-gui. The openVIM offers a northbound interface, based on REST (openVIM Application Programming Interface [API]), where enhanced cloud services (e.g. creation, deletion and management of instances, etc.) are offered. OpenNFV [20] is an orchestrator developed by HP to promote the development and evolution of NFV components and SDN infrastructure across various open source infrastructures. CloudNFV [21] is an orchestrator that implements NFV-based on cloud computing and SDN in a multi-vendor infrastructures. CloudNFV consists of three components, namely, the orchestrator, manager, and an active virtualization to manage existing network resources as well as maintaining an information base of the running services [19]. Cloudify [22] is a Topology and Orchestration Specification for Cloud Applications (TOSCA)-based cloud orchestrator that offers computing, network, and storage resources. It uses the Infrastructure as a Service (IaaS) API to provide a complete solution for automating and controlling application deployment and DevOps activities in a multi-cloud environment. Table 9.2 provides a summary of open source platforms that enable the 5G networks slicing. Some of the proof-of-concepts for network slicing based on the above mentioned orchestrators include multi-tenant hybrid slicing [25], RAN slicing [26] and 5G edge resource slicing [27], etc. Raza et al. [28] demonstrate an SDN/NFV-based orchestrator that enables resource sharing among different tenants and maximizes the profit of infrastructure providers based on a dynamic slicing approach using big data analytics. Capitani et al. [29] demonstrate the deployment of a 5G mobile network slice through the 5G-Transformer architecture experimentally.

Standardization efforts are being made regarding network slicing in different fora (e.g. Broadband Forum [BBF]), telecommunication industry (the Global System for Mobile Communications [GSMA], Open Networking Foundation [ONF] and NGMN) and by bodies including the ETSI, 3GPP, ITU, and IETF. The main goal has been to develop network slicing use cases and business models for vertical industries, service providers, and other potential 5G network slicing operators. The ETSI Zero touch network and Service Management Industry Specification Group (ZSM ISG) has been working on defining E2E network and service management requirements and developing architectures for multi-tenant and multi-domain environments in 5G infrastructure. The working groups define use cases and requirements, and architectures that provide security capabilities of E2E network slicing.

The ITU-T SG13 [30] standardization activities focus on developing the requirements and network management and orchestration framework that supports different industry verticals. The ITU-T SG15 defines a network slicing requirements for a transport network with SDN and developed a Slicing Packet Network (SPN) architecture for 5G transport. The IETF is working on developing the 5G network slicing architecture, network slice management, and orchestration techniques, including life-cycle management to coordinate E2E and domain orchestration [19] (Figure 9.3).

Table 9.2 Summary of orchestration enabling technologies in 5G networks slicing [23].

Orchestrator	Technology	Objectives	Technology Features	Management Features
OpenMANO	SDN, NFV	To provide a practical implementation of the NFV MANO reference architecture	OpenMANO, OpenVIM, REST API	—
OSM	Cloud networks and services	MANO with SDN control, multi-site/multi-VIM capability	OpenMANO, OpenVIM, JuJu, OpenStack	—
OPNFV	NFV	To facilitate the development of multi-vendor NFV solutions across various open source ecosystems	OpenStack, OpenDaylight	—
ECOMP	SDN/NFV, cloud and legacy networks	Software centric network capabilities and automated E2E services	TOSCA, YANG, OpenStack, REST-API	Improved OSS/BSS, service chain, policy management
T-NOVA	Network services and virtual resources	Network function as a service	OpenStack, OpenDaylight	OSS/BSS, service lifecycle
OpenBaton [24]	Heterogeneous virtual infrastructures	Enables virtual network services on a modular architecture	TOSCA, YANG, OpenStack, Zabbix	Event management and auto-scaling
Cloudify	NFV, cloud	To automate and deploy network services data centers in a multi-cloud environment	TOSCA, OpenStack, Docker, Kubernetes	Service chaining, OSS/BSS
ZOOM	NFV and cloud services	Optimization and monitoring of Network Functions-as-a-Service (NFaaS)	—	Improved OSS/BSS
CloudNFV	SDN/NFV-enabled cloud services	To enable the deployment of NFV in a cloud environment	OpenStack, TM Forum SID	Service chaining and OSS/BSS
HP OpenNFV	NFV	To allocate network resources based on global resource management policies	Helion OpenStack	—
Intel ONP	SDN and NFV	Accelerates the adoption of SDN and NFV in telecom, enterprise, and cloud markets	OpenStack, OpenDaylight	—

Source: Adapted from Taleb et al. [23].

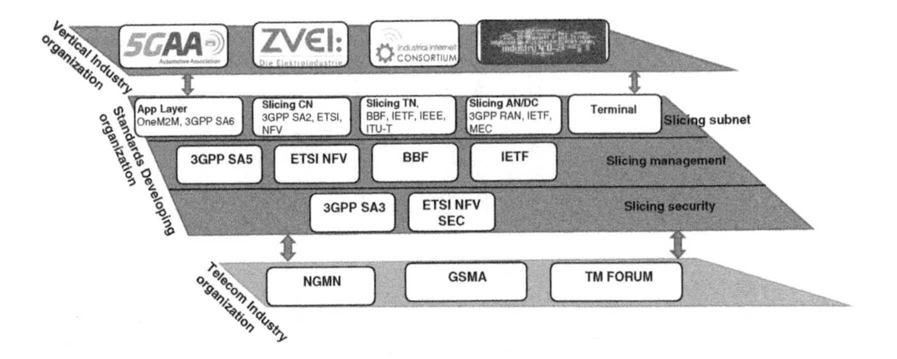

Figure 9.3 Network slicing relevant industry groups and SDOs landscape.

9.2 5G Network Slicing Orchestration and Management in Multi-domain

Multi-domain implementation in 5G networks is designed to allow numerous administrative domains at different levels to interface with different service and infrastructure providers. That way, multi-domain enables service demands from various domains are mapped into multi-operator and multi-technology domains, with each service Experience Level Agreement (ELA) need being met [31]. Recent works regarding multi-domain management and orchestration of resources in 5G network slicing include [31–34]. Wang et al. [32] present a QoE-driven 5G network slicing architecture in SDN/NFV-enabled 5G networks, with a focus on cognitive network control and management for E2E slicing operation as well as slice-based/enabled services across different operator domains. Guerzoni et al. [31] present a functional architecture for E2E management and orchestration in multi-domain 5G systems. The authors design and implement the MdO prototype that can automate network service provisioning over multiple-technologies that spans across multiple operators [6]. As to the vision of 5G!PAGODA, the authors in [35] introduce a 5G network slicing architecture that can create on-demand the slices of virtual mobile networks and customize them based on the changing needs of mobile services from different administrative domains. NESMO [33] describes a network slicing management and orchestration framework that extends the reference management architecture for network slicing introduced by the 3GPP [36]. NESMO provides an automated E2E network slice design, deployment and configuration in multiple network infrastructure resource domains that targets mobile network. Vaishnavi et al. [37] provide an experimental implementation of multi-domain orchestration where multi-operator services can be deployed and monitor the service for ELA/SLA compliance over 5G networks [19]. The authors in [38]

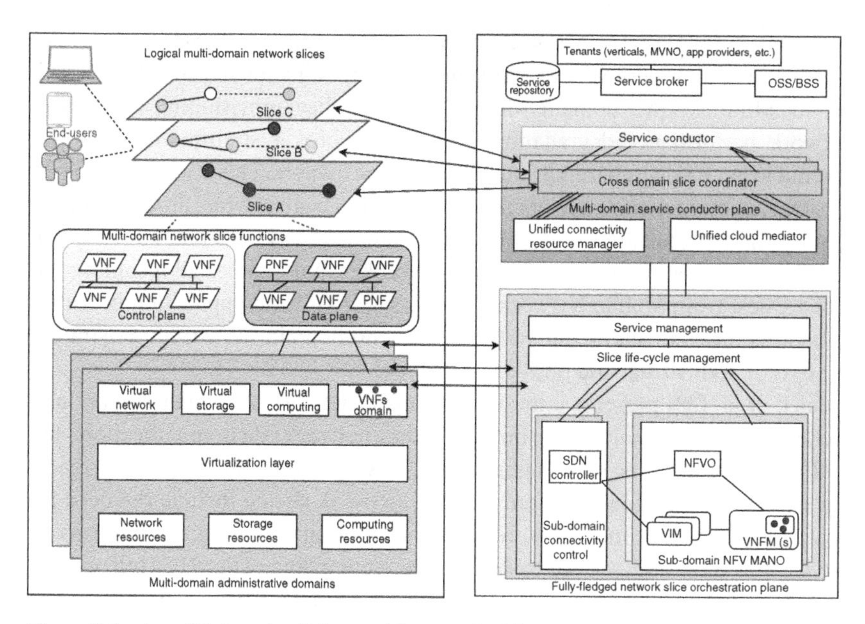

Figure 9.4 A multi-domain slicing architecture in 5G networks.

present a multi-domain management and orchestration architecture shown in Figure 9.4 that can deliver services and applications across federated domains using different components namely: (i) Domain-specific Fully-Fledged Orchestration Stratum, (ii) multi-domain Service Conductor Stratum, (iii) Sub-Domain MANO and Connectivity Stratum, and (iv) Logical Multi-Domain Slice Instance stratum. The domain-specific fully-fledged orchestration stratum uses the service management, sub-domain NFV-MANO, sub-domain SDN controller, and slice life-cycle management functions to distribute internal domain resources. That way, it establishes a federated network slice instance and provide the necessary Life-Cycle Management (LCM). The service management function is used for analyzing the requests from network slices as received from the cross-domain slice coordinator. It also identifies the Radio Access Network (RAN) and core network functions including value-added services. The sub-domain NFV-MANO function identifies the appropriate template for a network slice that is associated with a network catalog. It then forms a logical network graph and maps it to the underlying computing, storage and networking resources corresponding to a technology-specific slate [38]. The sub-domain SDN controller enables service function chaining for the allocated VNFs connecting to the remote cloud environments using PNFs. It also provides the network connectivity to the VNFs.

The multi-domain service conductor stratum maps service requirements of various multi-domain requests to their respective administrative domains by using two modules: the Cross-domain Slice Coordinator (CSC) and Service Conductor which performs network resources re-adjustments in different federated administrative domains during service policy provisioning changes or network performance degradation. The service broker stratum manages NSI revenues that involve charging and billing of slice owners. This is achieved by performing negotiation and network slicing admission control while different service requests arise from various administrative domains. It is worth mentioning that the service broker stratum acts as a service broker that handles all incoming service request from Mobile Network Operators (MNOs) and application providers as well as different industrial verticals. The sub-domain infrastructure stratum consists of both virtual and physical resources.

Taleb et al. [38] presents a the multi-domain network slice management and orchestration approaches for multi-domain network slice modification and multi-domain network slice configuration. The fundamental challenges regarding service management and control in 5G network slicing include the need for service management interfaces, resource isolation and sharing mechanisms, & service profiling and service-based management plane. The RESTfull models such as L3SM/L2SM [39, 40] and NFVIFA Os-Ma-Nfvo[2] have been considered to facilitate the information exchange and service management. These models are utilized for programmability to provide control capabilities among various administrative and technological domains, as well as third parties like industry verticals. However, the lack of robustness and performance assessment capabilities, as well as multi-domain connectivity and control considerations on federated resources, are flaws in these architectures. This calls for the development of new data models that can map and analyzes service requirements of the corresponding slice into the relevant cloud and networking resources. Efficient management approaches are proposed to perform slice admission control decisions and resource reservation across different administrative network domains [41]. The authors in [41] introduces an E2E slices architecture that exploits feedback information from mobile network slices to make orchestration decisions via a hierarchical control plane. However, service profiling algorithms for optimizing the mapping of allocated resources are acutely needed in future 5G network slicing environments [38].

9.2.1 Reference Architecture for Data-Driven Network Management in 6G and Beyond Networks

The network enabling technologies such as SDN, NFV, MEC, and AI/ML leverage the development and deployment of data-driven network management in 6G and beyond networks. Figure 9.5 shows the reference architecture

2 5ETSI GS NFV-IFA, Os-Ma-Nfvo reference point, interface, and information model specification.

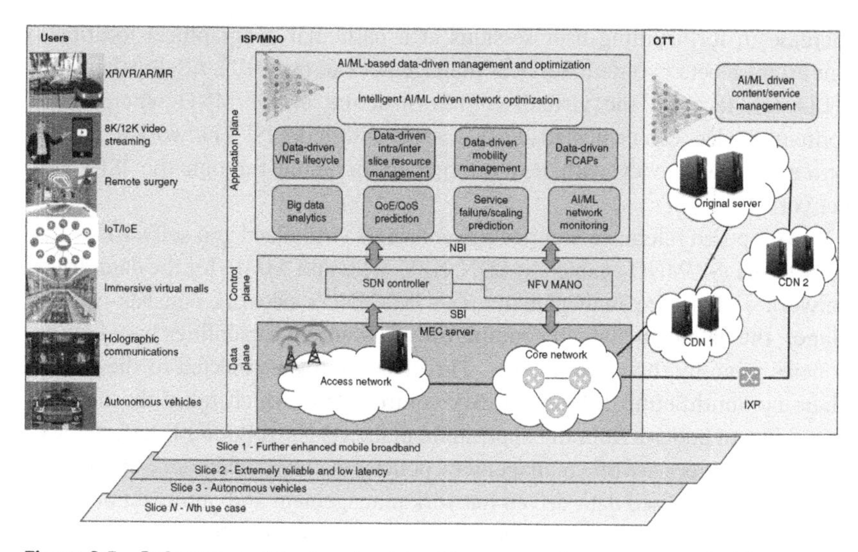

Figure 9.5 Reference architecture for data-driven network management in 6G and beyond networks. Source: Benjamín Núñez González/Wikimedia Commons/CC BY-SA 4.0; alan9187/Pixabay; PhotoMIX-Company/Pixabay.

for the data-driven network management in 6G and beyond using network enablers to support various use cases such as VR/AR/MR, remote surgery, autonomous vehicles, IoT/IoE, immersive virtual malls, 8K/12 video streaming, and holographic communications. The reference architecture considers the three main players in the E2E service delivery over the Internet: OTT service provider who delivers its service over the Internet, Internet Service Provider (ISP)/MNO provider who provides Internet connectivity to the users to use services/applications over the Internet and users who utilize the services delivered over the Internet.

The ISP/MNO's core networks are connected to OTT Autonomous System (AS) by either direct peering with Content Delivery Network (CDN) (highlighted by CDN 1 connection with ISP/MNO's core network in Figure 9.5) or by indirect peering through Internet eXchange Point (IXP) (represented by CDN 2 connection with the core network via IXP in Figure 9.5). Within OTT's AS, the content/services are distributed/deployed from the original server to different CDNs situated at multiple geographical locations using a cloud management system. The distribution of contents/services across OTT's AS delivers low latency for the service/content retrieval. In particular, the emerging next-generation Internet applications such as autonomous vehicles IoT/Internet of Everything (IoE) require extremely low E2E latency, which requires effective management of the contents/services on the OTT side where data-driven approaches can improve the service. For example, the AI-based data-driven approaches for predicting an increase in forthcoming user sessions at a particular geographical location(s) can promise better content/services management across OTT's AS. Furthermore, OTTs can leverage the virtualized infrastructure of ISP/MNO where OTT's content/services can be hosted at MEC server within the ISP's network to provide extremely low latency with less Internet traffic going through the ISP's core network.

The proposed reference architecture considers virtualized and softwarized network of the ISP/MNO enabled by SDN, NFV, MEC and AI/ML for the data-driven network and service management. The ISP/MNO's network consists of three planes: the data plane which contains data forwarding capabilities based on the policies given by the control plane. The data plane is connected to the control plane by SouthBound Interface (SBI); control plane which takes the network management policies from the application plane via NorthBound Interface (NBI) and implements network management policies network-wide and; application plane where AI-based data-driven network management and optimization application/services are deployed. The intelligent AI-driven network optimization can be deployed on top of the NFV MANO or SDN controller. Inspired by the O-RAN architecture, the AI-based data-driven network management application can also be deployed on top of near real-time RAN Intelligent Controller (RIC) or non-real-time RIC.

The AI-based data-driven network optimization utilizes big data analytics, AI/ML-based QoE/QoS prediction of emerging Internet applications in combination with AI-driven network monitoring and service failure/scaling prediction to perform automation of data-driven network management operations which includes data-driven FCAPS, data-driven mobility management for extremely reliable and low latency applications and data-driven intra/inter slice resource management.

The data-driven VNFs life-cycle entity is responsible for performing the life-cycle management of individual NFs and 6G mobile network instances. The data-driven inter/intra-slice resource management performs the overall allocation of both physical and virtual resources connectivity and networking characteristics within or outside the slice. It also provides self-management capabilities by focusing on self-optimization and self-healing of network slice instances during their deployment (life-cycle management) and operations. The data-driven mobility management component is designed to offer Mobility Management as a Service (MMaaS) where the SDN controller can provide efficient computational resource allocation to UEs and enable global optimization for managing user mobility in 6G and beyond networks. The data-driven FCAPS identifies and locates the faults occurring during the network operations and initiates remedial measures. It also ensures security is maintained in all layers of the data-driven architecture and provides accountability for all actions that affect the 6G and beyond networks. Big data analytics and AI/ML will play a central role in analyzing network utilization and traffic data patterns as well as supporting mission-critical applications such as remote surgery, holographic communications and public safety in real-time. The AI/ML monitoring and QoE/QoS prediction components are responsible for monitoring resource utilization, networks performance/status, and the overall QoE/QoS prediction based on user's demands and usage preferences in 6G and beyond networks.

The ISP/MNO's network comprises multiple network slices using SDN and NFV to ensure E2E service quality for the emerging Internet application use case. For example, the dedicated slice for Future Enhanced Mobile Broadband (FeMBB) applications such as 8K/12K video streaming provides E2E QoE/QoS. Similarly, different QoE/QoS requirements based multi-tenancy networks can be provided according to Service Level Agreements (SLAs). The intelligent AI-based data-driven network management approaches can be deployed on top of an SDN controller or NFV MANO as an application.

9.3 Conclusion

This chapter provides collaborative research projects and open source orchestrators, as well as standardization efforts for 5G network slicing. It is clear that network slice orchestrators have been designed and implemented in recent years by industry and academia primarily to facilitate resource orchestration and management in future 5G and beyond networks. Supporting the development and evolution of NFV components and softwarized infrastructure across multiple domains of 5G ecosystems has been the main driver of orchestrators. The advancement of developing NFV products and 5G services from projects such as OPNFV that focuses on multi-vendor NFV components ensure that performance targets and compatibility are met. CloudNFV, Cloudfy, and T-NOVA has been implementing NFV-based on cloud computing and SDN technologies in a multi-vendor environment to enable automated provisioning, configuration, monitoring and efficient operations of NFaaS on top of softwarized and virtualized 5G and beyond systems. Multi-service and multi-tenancy are concepts that have been explored in a number of 5G slicing research projects. 5GPPP initiatives such as 5G-NORMA and SESAME have been developing mechanisms regarding RAN multi-tenancy. The 5G-Crosshaul project as part of the 5GPP initiatives focuses on transport network characteristics directly relevant to combined fronthaul and backhaul, including per-tenant services combining computation, storage, switching, and transmission resource management.

Bibliography

1 Nightingale, J., Wang, Q., Alcaraz Calero, J.M. et al. (2016). QoE-driven, energy-aware video adaptation in 5G networks: the SELFNET self-optimisation use case. *International Journal of Distributed Sensor Networks* 2016. 1–15.

2 Dryjanski, M. (2017). Towards 5G- research activities in Europe (HORIZON 2020 projects. http://grandmetric.com/blog/2016/03/10/towards-5g-research-activities-in-europe-part-2-h2020-projects/ (accessed 3 May 2022).

3 Gouvas, P., Zafeiropoulos, A., Vassilakis, C. et al. (2017). Design, development and orchestration of 5G-ready applications over sliced programmable infrastructure. *1st International Workshop on Softwarized Infrastructures for 5G and Fog Computing (Soft5G)*, pp. 13–18, December 2017.

4 Bagaa, M., Taleb, T., Laghrissi, A. et al. (2018). Coalitional game for the creation of efficient virtual core network slices in 5G mobile systems. *IEEE Journal on Selected Areas in Communications* 36 (3): 1–15.

5 (2018). SliceNet: E2E Cognitive Network Slicing and Slice Management in 5G Networks. *IEEE International Symposium on Broadband Multimedia Systems and Broadcasting (BMSB)*, June 2018.

6 Sgambelluri, A., Tusa, F., Toka, L., et al. (2017). Orchestration of network services across multiple operators: the 5G exchange prototype. *European Conference on Networks and Communications (EuCNC)*, pp. 1–5, June 2017.

7 Barakabitze, A.A., Barman, N., Ahmad, A. et al. (2019). QoE management of multimedia services in future networks: a tutorial and survey. *IEEE Communication Surveys and Tutorials.* 22 (1): 526–565.

8 Galis, A., Valocchi, D., Clayman, S. et al. (2018). DevOps for 5G Network Function Virtualization chapter in the book “5G Networks: an European Vision”, August 2018.

9 Meredith, J.M. (2017). Management and orchestration of networks and network slicing; 5G Core Network (5GC) Network Resource Model (NRM) -Stage 1, *3GPP Technical specification (TS)*, June 2017.

10 Khatibi, S., Arnold, P., Gramaglia, M. et al. (2018). 5G-MoNArch: Deliverable D2.2: Initial overall architecture and concepts for enabling innovations. *3GPP Technical specification (TS)*, July 2018.

11 Silva, F.S.D., Lemos, M.O.O., Medeiros, A. et al. (2018). NECOS project: towards lightweight slicing of Cloud-federated infrastructures. *IEEE Conference on Network Softwarization (NETSOFT)*, June 2018.

12 Gramaglia, M., Digon, I., Friderikos, V. et al. (2016). Flexible connectivity and QoE/QoS management for 5G networks: the 5G NORMA. *IEEE International Conference on Communications Workshops (ICC)*, July 2016.

13 Li, X., Casellas, R., Landi, G. et al. (2017). 5G-Crosshaul network slicing: enabling multi-tenancy in mobile transport networks. *IEEE Communications Magazine* 55 (8): 128–137.

14 Xu, L., Assem, H., Buda, T.S. et al. (2013). CogNet: A network management architecture featuring cognitive capabilities. *European Conference on Networks and Communications (EuCNC)*, pp. 325–329, June 2013.

15 Costa-Perez, X., Garcia-Saavedra, A., Li, X. et al. (2017). 5G-Crosshaul: an SDN/NFV integrated Fronthaul/Backhaul transport network architecture. *IEEE Communications Magazine* 21 (1): 38–45.

16 Iovanna, P., Pepe, T., Guerrero, C. et al. (2018). 5G mobile transport and computing platform for verticals. *IEEE Wireless Communications and Networking Conference Workshops (WCNCW)*, pp. 266–271, April 2018.

17 Antevski, K., Martín-Pérez, J., Molner, N. et al. (2018). Resource orchestration of 5G transport networks for vertical industries. *Workshop 5G Cell-Less Nets*, September 2018.

18 Guerzoni, R., Vaishnavi, I., Perez Caparros, D. et al. (2017). Analysis of end-to-end multi-domain management and orchestration frameworks for software defined infrastructures: an architectural survey. *Transactions on Emerging Telecommunications Technologies* 28 (4): 1–19.

19 Barakabitze, A.A., Ahmad, A., Mijumbi, R., and Hines, A. (2020). 5G network slicing using SDN and NFV: a survey of taxonomy, architectures and future challenges. *Computer Networks* 167: 1–40.

20 HP (2015). HP OpenNFV Reference Architecture. http://www8.hp.com/us/en/cloud/nfv-overview.html? (Online: accessed 06 March 2018).

21 CloudNFV (2015). http://www.cloudnfv.com/ (Online: accessed 06 March 2018).

22 Cloudify (1978). https://cloudify.co/ (Online: accessed 10 April 2019).

23 Taleb, T., Samdanis, K., Mada, B. et al. (2017). On multi-access edge computing: a survey of the emerging 5g network edge cloud architecture and orchestration. *IEEE Communication Surveys and Tutorials* 19 (3): 1657–1681.

24 Carella, G.A., Pauls, M., Magedanz, T. et al. (2017). Prototyping NFV-based multi-access edge computing in 5G ready networks with open baton. *IEEE Conference on Network Softwarization (NetSoft)*, July 2017.

25 Guo, Q., Gu, R., Cen, M. et al. (2018). Multi-tenant hybrid slicing with cross-layer heterogeneous resource coordination in 5G transport network. *Optical Fiber Communication Conference*, March 2018.

26 Koutlia, K., Umbert, A., Garcia, S., and Casadevall, F. (2017). RAN slicing for multi-tenancy support in a WLAN scenario. *IEEE Conference on Network Softwarization (NetSoft)*, July 2017.

27 Boubendir, A., Guillemin, F., Le Toquin, C. et al. (2018). 5G edge resource federation: dynamic and cross-domain network slice deployment. *IEEE International Conference on Network Softwarization (NetSoft 2018) - Demo Sessions*, pp. 338–340, June 2018.

28 Raza, M.R., Natalino, C., Vidal, A. et al. (2018). Demonstration of resource orchestration using big data analytics for dynamic slicing in 5G networks. *European Conference on Optical Communication (ECOC)*, September 2018.

29 Capitani, M., Giannone, F., Fichera, S. et al. (2018). Experimental demonstration of a 5G network slice deployment through the 5G-transformer architecture. *European Conference on Optical Communication (ECOC)*, September 2018.

30 ITU-T (2018). Progress of 5G studies in ITU-T: overview of SG13 standardization activities. https://www.itu.int/en/ITU-T/Workshops-and-Seminars/20180604/Documents/Session1.pdf [Online: accessed 20 February 2019).

31 Guerzoni, R., Perez Caparros, D., Monti, P. et al. (2016). Orchestration of network services across multiple operators: the 5G exchange prototype. *European Conference on Networks and Communication*, June 2016.

32 Wang, Q., Alcaraz-Calero, J., Weiss, M.B., and Gavras, A. (2018). SliceNet: End-to-End cognitive network slicing and slice management framework in virtualised multi-domain, multi-tenant 5G networks. *IEEE International Symposium on Broadband Multimedia Systems and Broadcasting (BMSB)*, June 2018.

33 Devlic, A., Hamidian, A., Liang, D. et al. (2018). NESMO: Network slicing management and orchestration framework. *IEEE International Conference on Communications Workshops*, May 2018.

34 Draxler, S., Karl, H., Kouchaksaraei, H.R. et al. (2018). 5G OS: Control and orchestration of services on multi-domain heterogeneous 5G infrastructures. *European Conference on Networks and Communications (EuCNC): Network Softwarisation (NET)*, pp. 187–191, June 2018.

35 Ksentini, A., Afolabi, I., Bagaa, M. et al. (2017). Towards 5G network slicing over multiple-domains. *IEICE Transactions on Communications* E100-B (11): 1992–2006.

36 Kim, J., Kim, D., and Choi, S. (2017). 3GPP SA2 architecture and functions for 5G mobile communication system. *ICT Express* 3 (1): 1–8.

37 Vaishnavi, I., Czentye, J., Gharbaoui, M. et al. (2018). Realizing services and slices across multiple operator domains. *IEEE/IFIP Network Operations and Management Symposium*, April 2018.

38 Taleb, T., Afolabi, I., Samdanis, K., and Yousaf, F.Z. (2019). On multi-domain network slicing orchestration architecture & federated resource control. http://mosaic-lab.org/uploads/papers/3f772f2d-9e0f-4329-9298-aae4ef8ded65.pdf (accessed 3 May 2022).

39 Wen, B., Fioccola, G., Xie, C., and Jalil, L. (2018). A YANG Data Model for Layer 2 Virtual Private Network (L2VPN) Service Delivery, IETF RFC 8466, October 2018.

40 Wu, Q., Litkowski, S., Ogaki, K., and Tomotaki, L. (2018). A YANG Data Model for Layer 2 Virtual Private Network (L2VPN) Service Delivery, IETF RFC 8466, January 2018.

41 Salvat, J.X., Zanzi, L., Garcia-Saavedra, A. et al. (2018). Overbooking network slices through yield-driven end-to-end orchestration. *CoNEXT*, pp. 353–365, August 2018.

10

QoE Management of Multimedia Service Challenges in 5G Networks

This chapter provides future challenges for managing multimedia services focusing on the following important areas: emerging multimedia services, QoE-oriented business models in future softwarized network, intelligent QoE-based big data strategies in SDN/NFV, scalability, resilience, and optimization in SDN and NFV, multimedia communications in the Internet of Things (IoTs), and OTT-ISP collaborative service management in softwarized networks. The chapter provides various opportunities that need extensive research from the academia and industry regarding QoE-driven virtualized multimedia 3D services delivery schemes over 6G architecture, 3D cloud/edge models for multimedia services, and elastic 3D service customization via network intelligentization and QoE management and orchestration challenges over 6G networks.

10.1 QoE Management and Orchestration in Future Networks

Network softwarization and intelligentization with technologies such as SDN and NFV will continue to be key in 2030 networks [1, 2]. For decades, automated management has been the focus of network management research from academia and industry to enable autonomous monitoring of streaming status and network states, making control decisions, analyzing potential issues in the network, and executing corrective actions dynamically. Autonomic networking, knowledge-driven networks, policy-based management, and recently self-driving networks are some of attempts that have been made to achieve self-managing 5G networks [3, 4]. However, practical deployments in 5G systems have largely remained unrealized because many telecommunication and service provider stakeholders have different goals in terms of QoE control and management [3, 5].

Multimedia Streaming in SDN/NFV and 5G Networks: Machine Learning for Managing Big Data Streaming, First Edition. Alcardo Barakabitze and Andrew Hines.

Several projects have been initiated to resolve the network automation issues in 5G and beyond networks including AT&T's ECOMP [6], OSM [7] and ONAP project [8] to provide emphasis on life-cycle management of multimedia services over softwarized networks. Open source platforms (OpenMANO) [9], RIFT.ware [10], and JUJU [11] have been implementing different architectures to support the functionalities for management and orchestration of resources in future networks. Mayoral et al. [12] propose an architecture that enables dynamic resource allocation for interconnected virtual instances in distributed cloud locations. Boutaba et al. [13] present comprehensive learning paradigms and ML techniques applied to fundamental problems in networking, including resource and fault management, QoE-routing and classification, QoS and QoE management, traffic prediction, congestion control, and network security. Despite these efforts, the fundamental challenge is developing QoE-driven multiparty, multidomain, multitenant solutions which are crucial aspects in 6G and beyond networks for predicting demand and dynamically provisioning and reprovisioning resources to the end users. This calls for novel QoE prediction models and intelligent network/service management and orchestration strategies over softwarized 6G networks [3]. Extensive research from academia and industry is required to move from existing methods in use today for QoE delivery to the end users. The new QoE research in 6G and beyond networks should embrace the parameters from the user, context, system, and content pillars for providing a better understanding of QoE in multimedia services [3].

10.2 Immersive Media Experience in Future Softwarized and Beyond Networks

We have seen a significant advancement of technologies for AR/VR and Mixed Reality (MR) which is grouped under the umbrella of XR [14]. In AR, virtual objects and information are overlaid on the real world by wearing a head-mounted display (HMD) and headphones. This experience enhances the real world with digital details such as text, images, and animation [15]. In a VR experience, users are fully immersed in a simulated digital environment through special glasses or the screen of a mobile device [16]. MR blends both VR and AR experience. Microsoft's HoloLens is a great example of MR. Today, many applications such as audio-visual entertainment, interactive digital storytelling, immersive theatres, omnidirectional video, education, and games are adopting XR technologies. Immersive media often make use of VRA/AR to create or augment a virtual context of use [15]. XR is the ultimate level of immersion. However, users care more about immersive experiences on their devices (e.g. when playing a video game, video conferencing, or watching a movie on a TV). For example, during

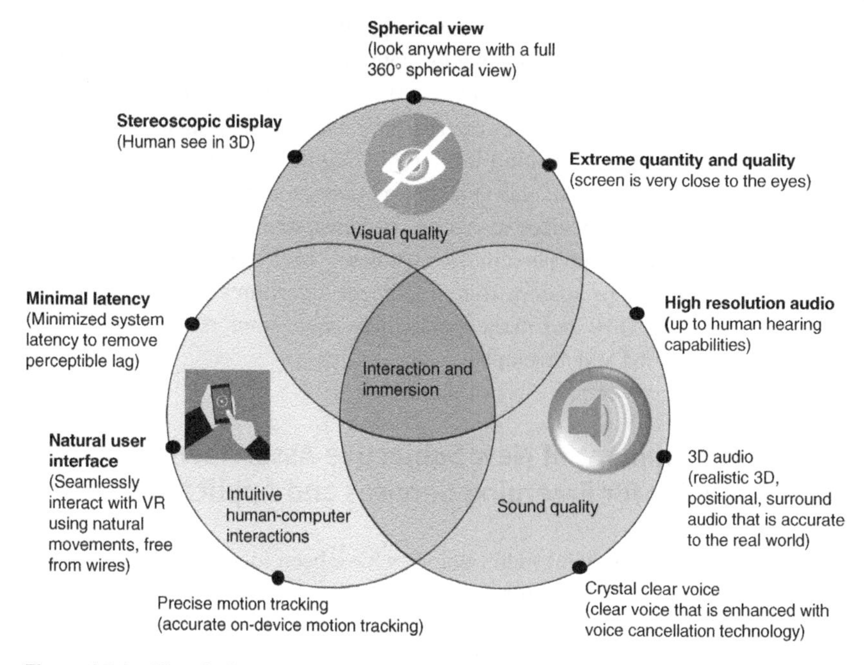

Figure 10.1 The challenge of immersive media experience in 6G and beyond networks. Source: Adapted from [16].

a live sporting event, an exotic vacation, or a concert performance, immersive experiences can stimulate a user's senses, vision, hearing, smell, taste, and imagination [16]. It is worth noting that truly immersive experiences in 6G and beyond systems will consider extreme requirements on several dimensions of the three pillars of immersion: visual quality, sound quality, and intuitive interactions as shown in Figure 10.1. However, due to the complexity and different constraints (performance, power, and cost) of devices in the era of 6G and beyond networks, achieving full immersion will be a very challenging task [17]. Teleimmersive (TI) applications (e.g. teleimmersive participatory media and immersive interaction spectating) that enable real time, multiparty interaction of users in a virtual world will be prevalent in the era of 6G and beyond networks [2]. For example, in a TI game scenario, multiple players from around the globe enter a 3D area constructed in real time. For the game players to have a good playing experience in the virtual world, a new ultralow latency-based TI architecture is needed that will utilize appropriate real-time 3D reconstruction methodologies and video compression standards in 6G and beyond networks [18].

To enhance and ensure smooth user experience, the QoS/QoE research should be the top priority in immersive media in 6G systems. The optimal approach of enhancing the key dimensions of sound quality, visual quality, and intuitive interactions is to utilize heterogeneous computing (parallel and distributed processing), cognitive technologies (ML/AI), and a new immersive 3D media platform that can optimize QoE metrics and monitor parameters from the 6G infrastructure and application-level [16]. Cognitive technologies such as computer vision and ML/AI can make experiences more immersive. These technologies can enable devices to reason, perceive, and take intuitive actions. These devices can learn and personalize customers' preference as well as enable intuitive interactions. To achieve this, intelligent algorithms that will improve and personalize the QoE and make experiences more immersive for the users have to be investigated and developed.

10.3 Development of New Subjective Assessment Methodologies for Emerging Services and Applications

Several subjective assessment methodologies have been identified for the omnidirectional videos using two common metrics (MOS and DMOS) [19]. Tran et al. [19] consider different parameters (e.g. video bitrate, resolution, content characteristics, and quantization parameter QP) to study their subjective perceptual quality effect for 360° video. Schatz et al. [20] conducted subjective research lab experiments to investigate the impact of stalling on QoE for 360° video streaming. The study conducted by Tran et al. [21] to investigate QoE for 360° videos concluded that video quality variations, delay, and interruptions can support an evaluation of the end user's QoE. The authors in [22] perform subjective quality evaluation comparison of H.265-encoded omnidirectional videos at different bit-rates for two different resolutions (FHD and UHD), while a testbed, a test method, and test metrics for QoE tests of viewport-adaptive streaming approaches for omnidirectional video are developed in [23]. The proposed methodology is validated using a tile-based streaming technology using video sequences, resolution, bandwidth, and network round-trip delay. However, the authors do not consider subjective quality measurements for 360° videos.

Although there are several subjective studies regarding 360° videos, the standardized methodologies for subjective metrics for 360° videos are still lacking [24]. This calls for efforts toward designing and developing quality assessment methods and QoE metrics for 360° videos. The research community should develop novel strategies that introduce less noise by adding new stitching, projection, and packaging formats. Viewport-dependent streaming and tile-based streaming are the commonly used techniques for implementing adaptive streaming in a VR environment for 360° video content, where a user can switch to neighboring views randomly during 360° video playback as provided in a recent work [24]. The only challenge is the facilitation of smooth viewport switching where a certain level of resilience has to be provided in order to remove a smooth viewport switching. ML can overcome this 360° video streaming challenge through objective and subjective QoE assessments. In this regard, deep reinforcement learning (DRL)-based approaches have to be developed for allocating video bitrates to the tiles in different regions of the 360° video frame.

10.4 QoE-aware Network Sharing and Slicing in Future Softwarized Networks

The concept of network slicing was introduced to provide an E2E logical network that runs on a common underlying (physical or virtual) infrastructure. A network slice is a mutually isolated entity with independent control and management capabilities [25]. 5G network slicing paradigm must be extended in the 6G network and beyond to enable data-driven virtualization, agility, reliability, and flexibility to support new use cases. Several technologies including haptic communication (zero-touch technology), industrial IoTs, use cases (human-centric services and 3D volumetric video streaming, extremely low-power communications (ELPC), Further-Enhanced Mobile BroadBand (FeMBB)), and others, will be supported in 6G. A new 6G network slicing data-driven architecture has to be developed to achieve the overall E2E network slice service management while supporting exploration of emerging use cases and scenarios in the future networks. Service management and orchestration of resources in multitenancy and multidomain environments in 6G networks and beyond is also an important area that still needs much investigation. It is worth mentioning that intelligent data-driven AI/ML mechanisms for 6G and beyond networks have to be developed to ensure that the service performance is maintained throughout the entire life cycle of a network slice across different administrative domains. Novel AI/ML approaches to perform a deep analysis and mapping of service requirements (resources allocation) of incoming multitenancy and multidomain requests onto the respective administrative domains in 6G networks and beyond have to be developed. Moreover, new security strategies to achieve security for interslice or intraslice networks or multioperator networks and the overall E2E security framework that can guarantee secured resource mapping across different 6G network administrative domains including the RAN, core, and transport networks are still unknown. For 6G network multioperator domains, we contemplate that an integrated fault-management, configuration, accounting, performance, and security (FCAPS) framework has to be developed to achieve a truly E2E service control, management, and orchestration of slice resources via a secured, interoperable, and reliable operations.

10.5 QoE Measurement, Modeling, and Testing Issues Over 6G Networks

Realistic and efficient measurements, modeling and testing models based on QoE for some 6G scenarios such as the new 3D massive MIMO channels [26] and high-mobility channels for enhancing and reducing delays in 6G are still a challenge and an open area requiring investigations [27]. With the increase of network nodes in 6G through the network densification (e.g. dense deployment of small cells) [28], new requirements for channel measurement/modeling in the spatial domain for supporting user mobility with advanced receivers capable of interference cancellation are crucial and should be considered in the design of 6G networks. In addition, subpath amplitude modulation, spherical wave modeling, and nonline-of-sight (NLOS) path-loss in mmWaves which are currently unknown also require major studies [29, 30].

Furthermore, the development of 6G-BSs and UEs which have an impact on user's QoE poses design and testing challenges due to the requirements of extremely high data rates, zero latency (<1 ms), wide channel bandwidth, complex antenna configurations, and the support for m-RATs. Some of the key challenges related to measurement and modeling of 6G propagation channels, testing of 6G, and beyond systems such as drone base-stations (drone-BS) and cellular-connected drone users (drone-UEs) are summarized in [27, 31].

10.6 QoE-Centric Business Models in Future Softwarized Network

SLAs are no longer sufficient as a means of establishing QoE-related contracts between service providers and customers. Web services have become a commonplace requirement for the formalization of new QoE-centric contracts between service providers and end users [25]. Unfortunately, most existing SLA modeling proposals are limited to technical aspects of QoS. Furthermore, they do not adhere to Web principles and a semantic approach when defining QoS for communication services utilizing QoE parameters [32]. ISPs in future telecommunication systems should focus on QoE marketization through ELAs that go beyond existing QoS-oriented SLAs to assure the Quality of Business (QoBiz) in the future communication ecosystem [25]. This will allow for the development of new business processes while ensuring a minimal level of QoE for end users. A business reference framework shown in Figure 10.2 can be a starting point for designing a new QoBiz model for service providers and customers to efficiently manage QoBiz in future networks.

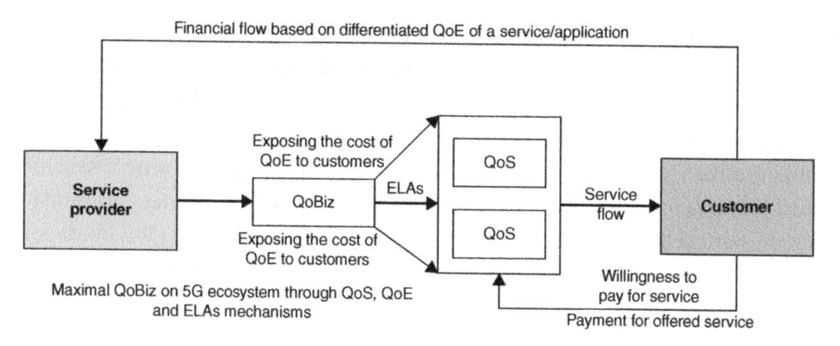

Figure 10.2 QoS, QoE, ELAs, and QoBiz relationship model for future communication systems.

The approach incorporates ELAs as a QoE-oriented augmentation beyond standard SLAs, which will enable new business provisioning to end users with QoE-differentiated services. It is worth mentioning that future SLAs should not be dependent on QoS, as they are in today's networks. Instead, they should be specified in terms of ELAs [33], which are QoE-only mechanisms for interfaces that require ELA definitions. This will improve and simplify the application/service planning process by simulating and modeling future softwarized system performance in terms of enhancing the end user's QoE. While user churn is still the most important factor for all businesses (e.g. MNOs, ISPs, OTTs) in the multimedia services industry, the relationship between QoE and user churn is an issue that requires further research. In [34], a mathematical function of the expected QoE given to the user is offered as an objective utility function of the user churn. The subjective validation of the user's QoE churn model, on the other hand, is still needed. In addition, future study on the impact of service price on customer satisfaction, perceived quality, and willingness to pay for services is also required. Similarly, Ahmad and Atzori [35] present a zero-rate QoE strategy for radio resource management in 5G networks while taking multimedia services into account. This paper proposes zero-rated QoE as an alternative to the widely utilized zero-rated data rate method, in which OTTP and MNO collaborate to provide QoE-aware services.

The collaborative management for service provisioning to customers by the ISPs and OTTP faces many challenges. The issues include the management and monitoring of the collaborative service based on the role played by of each entity at the interfaces for the information exchange. The optimization of OPEX and CAPEX in SDN- and NFV-based collaborative service management remains an open challenge [36]. Another issue is content distribution/duplication of OTTP content in the VSSs of the NFV-based ISP architecture, which is connected to cost of operation, load balancing, content replication, and content retrieval delay, particularly in the case of flash crowd appearance [25]. The OTTP-ISP collaborative QoE-aware service management, therefore, necessitates SLAs/ELAs and commercial agreements among the providers for service management policies. This will result in service supply to various users in different classes of service. In addition, algorithm for collaborative QoE-aware service management is also required, where several administrative domains of service management, such as QoE-fairness and business models, must be taken into account. The service's collaborative QoE management may also necessitate QoE prediction models that can predict longer-duration videos rather than shorter-duration videos.

10.7 Novel QoE-driven Virtualized Multimedia 3D Services Delivery Schemes Over 6G Networks

The perceived user experience today is mainly driven by service or application and management of networks [71]. The service and application management is within the control and power of content and service providers), while network management is within the control domain and power of ISPs and network operators [3]. These stakeholders are still isolated from each other (Table 10.1). However, in the era of 6G and beyond networks, this isolation should be raised for mutual benefits especially in the provision of multimedia 3D services. This calls for the development of innovative solutions that would enable the delivery of 3D services over 3D networks by looking at the service delivery chain as an ecosystem including end users, ISPs, network operators, content/service providers. The developed schemes should holistically extract intelligence from the network, applications, and servers and take smart decisions in improving the end user's QoE of the delivered 3D services. Novel orchestration and software-defined control 3D-based architectures, algorithms, and interfaces have to be designed, developed, and evaluated over 6G and beyond networks to meet the demands of critical multimedia 3D use-case scenarios.

10.8 Novel 3D Cloud/Edge Models for Multimedia Applications and Elastic 3D Service Customization

The research community should develop innovative, abstraction 3D cloud/edge models suitable for multimedia 3D services and applications such as Video Game as a 3D Service (VGaa3DS) or Video Streaming as a 3D Service (VSaa3DS). They should also address the challenge of 3D service customization over 6G architecture by shaping the underlying network resources according to the performance

Table 10.1 A summary of research directions and recommendations in future softwarized networks.

Research topic in SDN/NFV	Research challenges and recommendations
OTTP and ISP Collaboration [34, 37, 38]	The reference architecture, optimization algorithms for service management, and commercial models for collaborative QoE-aware service management solutions are required
HTTP adaptive streaming over MPTCP/QUIC [39–41], immersive video streaming	More investigation on the impact of MPTCP/QUIC and segment routing on adaptive streaming over future softwarized networks such as 6G is needed. For immersive video streaming, viewport-dependent solutions for VR streaming in future communication systems have to be extensively investigated
Strategies for managing big data [42–44]	Creating a QoE-based dynamic modeling procedure that leverage big data extracted over softwarized networks. Revisit the relationship between SDN, NFV, and big data in the context of future networks.
Scalability, resilience, and QoE optimization [45–49, 50–52]	QoE-aware multiobjective optimization models should be developed in future softwarized networks that consider SDN controller and VNFs/VNFMs placements. Novel QoE-aware dynamic resource management mechanisms that consider multidomain and distributed VNFs and network survivability in future softwarized networks are needed.
Network sharing and slicing [53–61]	To achieve the effectiveness of NS over software-defined/driven networks, the problems relating to the placement of NFs inside a slice, slice orchestration, or interdomain services, then network slicing must be investigated further. Isolation between slices, mobility management, dynamic slice formation, and security [62] are all aspects that require significant research.
QoE-centric business models [63, 64]	Developing new QoE business models. This is so because, SLAs used today are not appropriate as the means to provide QoE-contracts between service providers and customers in FNs.
Network performance, evaluation, and benchmarks [65–67]	Developing performance evaluation methodology and benchmarking tool (e.g. TRIANGLE [66] and Open5GCore [65]) that would help application developers and device manufacturers to test and benchmark new applications, devices, and services. It is important also to develop virtualization testbeds such as OpenSDNCore [67] that will provide practical implementation of the future network evolution paradigms leveraging NFV/SDN environment.
Security, privacy, and trust [68–70]	The impact of traffic encryption on users' QoE over softwarized networks is an urgent area that needs investigation. Same way, new personalized QoE models in FNs that include privacy and security together with classical video streaming metrics over SDN/NFV have to be developed.

requirements of incoming multimedia service requests. Similar to the approach used by the H2020 VITAL project, researchers should develop open interfaces to allow third parties and vertical segments to control and manage the underlying 6G network infrastructure. This would facilitate the capability of 3D networking architecture to form VNFs and 3D services in a flexible manner while considering the requirements of the desired service. Network intelligentization via ML/AI will provide the means of composing novel virtual functions for 3D services on-demand using modular and subfunctions strategies to address a full service customization in 6G networks. Furthermore, an intelligent orchestration system based on SDN/NFV that will allocate flexible VNFs, create network slices, and communicate with external entities in 6G networks has to be developed. To build an intelligent-driven 6G architecture, each external entity has to be equipped with sufficient computing, communication, and caching (3C) resources to support intelligent operations, control, management, and monitoring of 3D media services.

10.9 Security, Privacy, Trust

The 5G security mechanisms have been designed by integrating the 3GPP 4G security architecture. Security threats such as signaling plane, interoperator security, radio interfaces, user plane, bidding down, masquerading, privacy, and man-in-the-middle security issues have been considered in the architectural design of 5G networks. However, with the emergence of new use cases (high-precision communications, holographic telepresence, pervasive connectivity, etc.), current reactive security approaches in 4G/5G systems cannot be directly applied in 6G and beyond networks. New proactive security mechanisms should be developed to provide centralized security management in a shared 6G network. New quantum security techniques seem to be promising and can make 6G and beyond systems more secure which is different from the current public-key cryptography protocols.

More research efforts should be devoted to investigating new approaches regarding quantum security aspects in 6G and beyond networks. The current communication channels (data channel, control-channel, and intercontroller channel) in softwarized 5G networks are protected using Transport Layer Security (TLS) or Secure Sockets Layer (SSL) which lack strong authentication mechanisms [72]. The TLS and SSL are also vulnerable to Internet Protocol (IP) attacks and scanning attacks. This calls for new data-driven end-to-end security approaches across all layers (application, control/signaling, and data planes) based on AI/ML to secure the communication channels in softwarized and intelligentized 6G and beyond networks especially in the core networks which are envisaged to use multiple SDN controller interfaces. Novel ML-based NFV security strategies have to be developed in 6G networks to ensure that components (NFV MANO, virtual machines, hypervisors, VNFs) are protected from both virtual element-based attacks and physical attacks to maintain system reliability. The Blockchain technology will also have a significant impact in 6G spectrum sharing, distributed ledger technologies, and softwarized network decentralization.

New blockchain-based data-driven architecture for mobile service authorization and security improvements of media access control protocols have to be developed. Therefore, efficient decentralized attribute-based encryption strategies will be needed to handle security and privacy issues among billions of devices and thousands of cells deployed by operators. 6G networks will support even more sharing of network resources/infrastructures to reduce costs. Virtual networks which are shared by different operators may run and control multiple applications such as broadband, autonomous vehicles, and healthcare slices which require low latency. Handling data and information exchange among the operators in shared network infrastructures in future 6G and beyond networks will be a challenge. Therefore, new E2E data confidentiality mechanisms that will protect customers' and operators' privacy are required. Novel approaches for ownership/licensing of personal data and information between various stakeholders (mobile network operator, service providers, and third parties) in 6G and beyond systems have to be developed.

10.10 Conclusion

The limitation of 5G networks compared to what was promised during the last decade is pushing the multimedia research community to focus on bringing in new, innovative data-driven architecture with the vision toward 6G and beyond networks. Despite the recent efforts toward overcoming the control, monitoring, and management challenges of multimedia streaming services, and the rapid evolution of future softwarized networks that leverage SDN and NFV, there are still significant gaps in some areas that need extensive research and investigation. This chapter identifies some research opportunities for future explorations in the area of managing and controlling as well as orchestrating resources in the network, adaptive video streaming based on MPTCP/QUIC protocols, emerging multimedia applications and services such as immersive videos, holographic videos, 360° videos), and OTTP-ISP collaborative service management.

To keep up with the rate at which service providers and MNOs are proposing SDN and NFV for QoE provisioning to end users, more research into multimedia services should be performed in the following critical areas that have not been widely investigated in the past. QoE-aware dynamic control and management of video streaming services, QoE-centric business models in softwarized networks, QoE-based AI/ML and big data strategies, security, privacy and trust models, QoE-aware network sharing and slicing in future softwarized networks, development of QoE-based subjective and assessment models, immersive media experience in future softwarized, and beyond networks and network performance, evaluation, and benchmarks.

Bibliography

1 Nakamura, T. (2020). 5G evolution and 6G. *International Symposium on VLSI Design, Automation and Test (VLSI-DAT)*, September 2020.

2 Akyildiz, I.F., Kak, A., and Nie, S. (2020). 6G and beyond: the future of wireless communications systems. *IEEE Access* 8: 13399–134030.

3 Barakabitze, A.A., Barman, N., Ahmad, A. et al. (2019). QoE management of multimedia streaming services in future networks: a tutorial and survey. *IEEE Communication Surveys and Tutorials* 22 (1): 526–565.

4 5G Media: Programmable edge-to-cloud virtualization fabric for the 5G Media industry. http://www.5gmedia.eu/ (accessed 20 May 2021).

5 Barakabitze, A.A., Liyanage, M., and Hines, A. (2020). QoESoft: QoE management architecture for softwarized 5G networks. *IEEE International Conference on Communications Workshops (ICC Workshops)*, June 2020.

6 AT & T. Enhanced Control, Orchestration, Management, and Policy (ECOMP) Architecture. https://about.att.com/content/dam/snrdocs/ecomp.pdf (accessed 27 February 2018).

7 OSM ESTI NFV. https://osm.etsi.org/ (accessed 27 February 2018).

8 ONAP:Linux Foundation Project. https://www.open-o.org/ (accessed 27 February 2018).

9 Telefonica Research Institute. https://github.com/nfvlabs/openmano (accessed 27 February 2018).

10 RIFT.io. https://riftio.com/tag/rift-ware/ (accessed 27 February 2018).

11 Canonical. http://www.ubuntu.com/cloud/juju (accessed 27 February 2018).

12 Mayoral, A., Casellas, R., Muñoz, R. et al. (2017). Need for a transport API in 5G for global orchestration of cloud and networks through a virtualized infrastructure manager and planner. *Journal of Optical Communication Networks* 9 (1): A55–A62.

13 Boutaba, R., Salahuddin, M.A., Limam, N. et al. (2018). A comprehensive survey on machine learning for networking: evolution, applications and research opportunities. *Journal of Internet Services and Applications* 9 (16): 1–99.

14 Torres Vega, M., Christos L., Sergi A., et al. (2020). Immersive interconnected virtual and augmented reality: a 5G and IoT perspective. *Journal of Network and Systems Management* 28 (4): 796–826.

15 Perkis, A. and Timmerer, C. (2020). QUALINET white paper on definitions of immersive media experience (IMEx). *14th QUALINET Meeting (online)*, May 2020.

16 Driving the New Era of Immersive Experiences. https://www.qualcomm.com/media/documents/files/whitepaper-driving-the-new-era-of-immersive-experiences-qualcomm.pdf (accessed 13 August 2022).

17 Gul, S., Podborski, D., and Son, J. (2020). Cloud Rendering-based Volumetric Video Streaming System for Mixed Reality Services, pp. 1–4. *https://arxiv.org/abs/2003.02526*.

18 Caruso, G., Nucci, F.S., Rizou, S. et al. (2020). Embedding 5G solutions enabling new business scenarios in Media and Entertainment Industry. *IEEE 2nd 5G World Forum (5GWF)*, April 2020.

19 Tran, H.T. et al. (2017). A subjective study on QoE of 360 video for VR communication. *19th International Workshop on Multimedia Signal Processing (MMSP)*, pp. 1–6, October 2017.

20 Schatz, R. et al. (2017). Towards subjective quality of experience assessment for omnidirectional video streaming. *9th International Conference on Quality of Multimedia Experience (QoMEX)*, pp. 1–6, June 2017.

21 Tran, H.T.T. et al. (2016). A multi-factor QoE model for adaptive streaming over mobile networks. *IEEE Globecom Workshops (GC Wkshps)*, pp. 1–6, December 2016.

22 Singla, A. et al. (2017). Comparison of subjective quality evaluation for HEVC encoded omnidirectional videos at different bit-rates for UHD and FHD resolution. *Proceedings of the on Thematic Workshops of ACM Multimedia, Mountain View*, pp. 511–519, October 2017.

23 Singla, A. et al. (2019). Subjective quality evaluation of tile-based streaming for omnidirectional videos. *10th ACM Multimedia Systems Conference*, pp. 232–242, June 2019.

24 Shafi, R. et al. (2020). 360-degree video streaming: a survey of the state of the art. *Symmetry* 12 (9). 1–31.

25 Barakabitze, A.A., Barman, N., Ahmad, A. et al. (2020). QoE management of multimedia services in future networks: a tutorial and survey. *IEEE Communication Surveys and Tutorials*. 22 (1): 526–565.

26 Ademaj, F., Taranetz, M., and Rupp, M. (2016). 3GPP 3D MIMO channel model: a holistic implementation guideline for open source simulation tools. *EURASIP Journal on Wireless Communications and Networking* 55 1–14.

27 Medbo, J., Borner, K., Haneda, K. et al. (2019). Beyond 5G with UAVs: foundations of a 3D wireless cellular network. *IEEE Transactions on Wireless Communications* 18 (1): 357–372.

28 Bhushan, N., Li, J., Malladi, D. et al. (2014). Network densification: the dominant theme for wireless evolution into 5G. *IEEE Communications Magazine* 52 (2): 82–89.

29 Samimi, M.K., Sun, S., and Rappaport, T.S. (2016). MIMO channel modeling and capacity analysis for 5G millimeter-wave wireless systems. *European Conference on Antennas and Propagation (EuCAP'2016)*, pp. 1–5, April 2016.

30 Fadlullah, Z.M. and Kato, N. (2020). HCP: heterogeneous computing platform for federated learning based collaborative content caching towards 6G networks. *IEEE Access* 10 1.

31 Hossain, E. and Hasan, M. (2015). 5G cellular: key enabling technologies and research challenges. *IEEE Instrumentation and Measurement Magazine* 18 (3): 64–68.

32 Barakabitze, A.A., Ahmad, A., Mijumbi, R., and Hines, A. (2020). 5G network slicing using SDN and NFV: a survey of taxonomy, architectures and future challenges. *Computer Networks* 167: 1–40.

33 Varela, M., Zwickl, P., Reichl, P. et al. (2015). From service level agreements (SLA) to experience level agreements (ELA): the challenges of selling QoE to the user. *Proceedings of IEEE ICC QoE-FI*, June 2015.

34 Ahmad, A., Floris, A., and Atzori, L. (2016). QoE-aware service delivery: a joint-venture approach for content and network providers. *2016 Eighth International Conference on Quality of Multimedia Experience (QoMEX)*, pp. 1–6. IEEE.

35 Ahmad, A. and Atzori, L. (2020). MNO-OTT collaborative video streaming in 5G: the zero-rated QoE approach for quality and resource management. *IEEE Transactions on Network and Service Management* 17 (1): 361–374.

36 Ahmad, A. and Atzori, L. (2019). MNO-OTT collaborative video streaming in 5G: the zero-rated QoE approach for quality and resource management. *IEEE Transactions on Network and Service Management* 17 (1): 361–374.

37 Floris, A., Ahmad, A., and Atzori, L. (2018). QoE-aware OTT-ISP collaboration in service management: architecture and approaches. *Transactions on Multimedia Computing Communications and Applications*. 14 1–24.

38 Ahmad, A., Floris, A., and Atzori, L. (2017). OTT-ISP joint service management: a customer lifetime value based approach. *2017 IFIP/IEEE Symposium on Integrated Network and Service Management (IM)*, pp. 1017–1022. IEEE.

39 Barakabitze, A.A., Sun, L., Mkwawa, I.-H., and Ifeachor, E. (2018). QualitySDN: Improving video quality using MPTCP and segment routing in SDN/NFV. *IEEE Conference on Network Softwarization*, May 2018 (Submitted).

40 Barakabitze, A.A., Sun, L., Mkwawa, I.-H., and Ifeachor, E. (2018). A novel QoE-centric SDN-based multipath routing approach of mutimedia services over 5G networks. *IEEE International Conference on Communications*, May 2018.

41 Wu, J., Yuen, C., Cheng, B. et al. (2016). Streaming high-quality mobile video with multipath TCP in heterogeneous wireless networks. *IEEE Transactions on Mobile Computing* 15 (9): 2345–2361.

42 Cui, L., Yu, F.R., and Yan, Q. (2016). When big data meets software-defined networking: SDN for big data and big data for SDN. *IEEE Network* 30 (1): 58–65.

43 Lin, B.-S.P., Lin, F.J., and Tung, L.-P. (2016). The roles of 5G mobile broadband in the development of IoT, big data, cloud and SDN. *Communications and Network* 8: 9–21.

44 Barona López, L.I., Maestre Vidal, J., and García Villalba, L.J. (2017). An approach to data analysis in 5G networks. *Entropy* 19 (2): 1–23.

45 Bannour, F., Souihi, S., and Mellouk, A. (2017). Scalability and reliability aware SDN controller placement strategies. *13th International Conference on Network and Service Management (CNSM)*, November 2017.

46 ul Huque, M.T.I., Si, W., Jourjon, G., and Gramoli, V. (2014). Large-scale dynamic controller placement. *IEEE TNSM*, pp. 63–76, March 2014.

47 Ksentini, A., Bagaa, M., and Taleb, T. (2016). On using SDN in 5G: the controller placement problem. *IEEE Global Communications Conference (GLOBECOM)*, March 2016.

48 Wang, G., Zhao, Y., Huang, J., and Wang, W. (2017). The controller placement problem in software defined networking: a survey. *IEEE Network* 31 (5): 21–27.

49 Abdel-Rahman, M.J., Mazied, E.D.A., Teague, K. et al. (2017). Robust controller placement and assignment in software-defined cellular networks. *26th International Conference on Computer Communication and Networks (ICCCN)*, July 2017.

50 Yao, L., Hong, P., Zhang, W. et al. (2015). Controller placement and flow based dynamic management problem towards SDN. *IEEE International Conference on Communication Workshop (ICCW)*, June 2015.

51 Killi, B.P.R. and Rao, S.V. (2017). Capacitated next controller placement in software defined networks. *IEEE Transactions on Network and Service Management* 14 (3): 514–527.

52 Carpio, F., Jukan, A., and Pries, R. (2017). Balancing the Migration of Virtual Network Functions with Replications in Data Centers, pp. 1–8. *arXiv:1705.05573v2*.

53 Biczok, G., Dramitinos, M., Toka, L. et al. (2017). Manufactured by software: SDN-enabled multi-operator composite services with the 5G exchange. *IEEE Communications Magazine* 55 (4): 80–86.

54 Sgambelluri, A., Tusa, F., Toka, L. et al. (2017). Orchestration of network services across multiple operators: the 5G exchange prototype. *European Conference on Networks and Communications (EuCNC)*, pp. 1–5, June 2017.

55 Costa-Perez, X., Garcia-Saavedra, A., Li, X. et al. (2017). 5G-crosshaul: an SDN/NFV integrated fronthaul/backhaul transport network architecture. *IEEE Communications Magazine* 21 (1): 38–45.

56 Ordonez-Lucena, J., Ameigeiras, P., Lopez, D. et al. (2017). Network slicing for 5G with SDN/NFV: concepts, architectures, and challenges. *IEEE Communications Magazine* 55 (5): 80–87.

57 Chartsias, P.K., Amiras, A., Plevrakis, I. et al. (2017). SDN/NFV-based end to end network slicing for 5G multi-tenant networks. *European Conference on Networks and Communications (EuCNC)*, June 2017.

58 Rost, P., Mannweiler, C., Michalopoulos, D.S. et al. (2017). Network slicing to enable scalability and flexibility in 5G mobile networks. *IEEE Communications Magazine* 55 (5): 72–79.

59 Gramaglia, M., Digon, I., Friderikos, V. et al. (2016). Flexible connectivity and QoE/QoS management for 5G networks: the 5G NORMA. *IEEE International Conference on Communications Workshops (ICC)*, July 2016.

60 Crippa, M.R., Arnold, P., Friderikos, V. et al. (2017). Resource sharing for a 5G multi-tenant and multi-service architecture. *23th European Wireless Conference Proceedings of European Wireless*, pp. 1–6, May 2017.

61 Li, X., Casellas, R., Landi, G. et al. (2017). 5G-crosshaul network slicing: enabling multi-tenancy in mobile transport networks. *IEEE Communications Magazine* 55 (8): 128–137.

62 Li, X., Samaka, M., Chan, H.A. et al. (2013). Network slicing for 5G: challenges and opportunities. *IEEE Internet Computing* 21: 20–27.

63 Rivera, D. and Cavalli, A. (2016). QoE-driven service optimization aware of the business model. *30th International Conference on Advanced Information Networking and Applications Workshops (WAINA)*, pp. 725–730.

64 Rivera, D., Cavalli, A.R., Kushik, N., and Mallouli, W. (2016). An implementation of a QoE evaluation technique including business model parameters. *11th International Conference on Software Paradigm Trends*, pp. 138–145, August 2016.

65 Open5GCore. Fraunhofer FOKUS Research Institute. http://www.open5gcore.org/ (accessed 04 April 2018).

66 Cattoni, A.F., Madueño, G.C., Dieudonne, M. et al. (2016). An end-to-end testing ecosystem for 5G: the triangle testing house test bed. *European Conference on Networks and Communications (EuCNC*, IEEE.

67 OpenSDNCore. OpenSDNCore - Research & Testbed for the carrier-grade NFV/SDN environment. https://www.opensdncore.org// (accessed 17 September 2019).

68 ETSI (2014). ETSI GS NFV-SEC 003 V1.1.1: Network Functions Virtualisation (NFV); NFV security; Security and trust guidance, ETSI Ind. Spec. Group (ISG) Netw. Functions Virtualisation (NFV).

69 Scott-Hayward, S., Natarajan, S., and Sezer, S. (2015). A survey of security in software defined networks. *IEEE Communication Surveys and Tutorials* 18 (1): 624–654.

70 Orsolic, I., Pevec, D., Suznjevic, M., and Skorin-Kapov, L. (2017). A machine learning approach to classifying YouTube QoE based on encrypted network traffic. *Multimedia Tools and Applications* 76 (21): 22267–22301.

71 Schwarzmann, S. et al. (2019). Estimating video streaming QoE in the 5G architecture using machine learning. *Proceedings of the 4th Internet-QoE Workshop on QoE-based Analysis and Management of Data Communication Networks*, pp. 7–12, October 2019.

72 Khan, R., Kumar, P., Jayakody, D.N.K., and Liyanage, M. (2019). A survey on security and privacy of 5G technologies: potential solutions, recent advancements and future directions. *IEEE Commnication Surveys and Tutorials* 22 (1): 196–248.

11

Multimedia QoE-Driven Services Delivery Toward 6G and Beyond Network

The chapter builds on the achievements of 5G networks to provide the first roadmap toward 6G and beyond networks in terms of requirements, use-case scenarios and service classes (e.g. holographic and future media communications, human-centric services and three-dimensional (3D) volumetric video streaming, and new video compression standards). We also present potential technologies that will be dominant through 2025–2030 in shaping the vision of 6G and beyond networks. Furthermore, the chapter provides various features and technologies as well as challenges and research directions in 6G and beyond networks. The comprehensive goal of this chapter is to encourage researchers from the industry and academia to work together toward the realization of softwarized 6G and beyond networks while tackling the critical research challenges regarding management of emerging 3D multimedia services and applications.

11.1 The Roads Toward 6G and Beyond Networks

5G networks [1, 2] are offering several advancements regarding the network softwarization and virtualization, new frequency bands, massive Multiple-Input Multiple-Output (MIMO), massive IoTs, ultradensification etc. 5G use-cases such as enhanced mobile broadband (eMBB), ultrareliable low-latency communication (uRLLC), and massive machine-type communications (mMTC) support different applications and services including multimedia, VR/AR, Machine to Machine (M2M)/D2D communications, eXtended Reality (XR), eMBB. However, 5G is always associated with trade-offs because of the increasing societal needs, technological breakthroughs, and advancement of new services and applications toward 6G networks. The next generation of VAR (holographic teleportation) and XR experiences need microsecond-level latency and Terabytes Per Second (Tbps)-level data rates which cannot be sufficiently supported by 5G systems. The upcoming Industry X.0 paradigm also presents the next evolution of modern

Multimedia Streaming in SDN/NFV and 5G Networks: Machine Learning for Managing Big Data Streaming, First Edition. Alcardo Barakabitze and Andrew Hines.

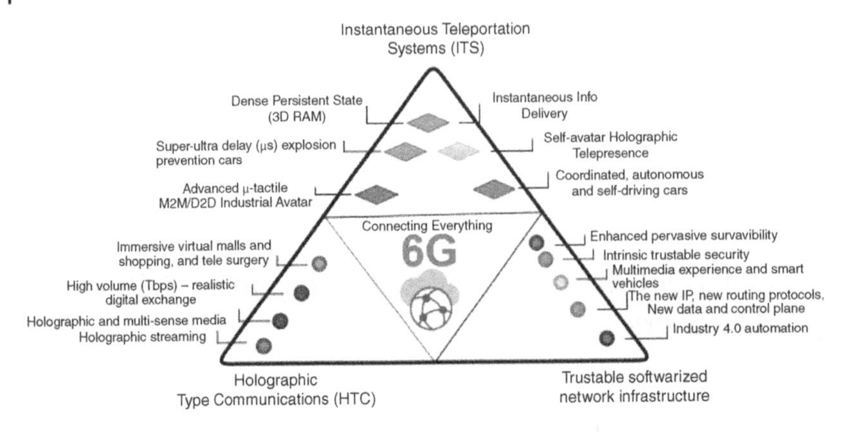

Figure 11.1 Future 6G communication services and applications.

industrial and networked factories that will need reliable high-throughput connectivity across thousands of devices often with sub-millisecond response times [3]. 5G is not designed to serve this connection density that is beyond the 10^6 km^2 metric. 5G systems will not be able to support the evolution of 8K/12K VR, volumetric video and 6 Degrees of Freedom (6DoF). As of today, 360° 8K and 360° 4K video require 50–200 and 10–50 Mbps, respectively, which 5G can offer. However, the future of free viewpoint video or 6DoF technologies which are critical for full immersive experiences need 200 Mbps to 5 Gbps. This puts 5G in a position to be challenged. The rapid development of data-centric and automated processes makes 6G to play a key role toward opening the roadmap in a way that 5G is fundamentally not designed to do.

The development of 6G [3–5] networks has already started from different consortia, companies, and governments. The network 2030 builds upon the achievements of International Mobile Telecommunications 2020 (IMT2020) to define capabilities of 2030 networks, emerging technologies, and corresponding communication services and applications. The IMT 2030 is investigating the needs to satisfy the requirements for the intelligent information society of 2030. The International Telecommunication Union (ITU) Telecommunication Standardization Sector (ITU-T) established the Focus Group Technologies for Network 2030 (FG NET-2030) within the ITU Study 13. The FG NET-2030 investigates the future 2030 network architecture and requirements support different use-cases such as (holographic type communications (HTC) and high-precision communication demands of emerging market verticals. Several projects have been initiated in the European Commission's Horizon 2020 to investigate the network capabilities beyond 5G (B5G) and 6G networks. Projects such as TERRANOVA aim to provide optical network QoE in B5G systems using

THz technologies. TERAPOD investigates the feasibility of ultrahigh bandwidth wireless access networks that operate in the terahertz band. The DEDICAT-6G [6] project develops mechanisms that can provide an intelligent placement of computation in mobile networks and maintain an efficient dynamic connectivity. The project also addresses security, privacy, and trust assurance challenges in 5G networks especially for mobile edge services. Moreover, the DEDICAT 6G designs and develops an architecture by transforming B5G networks into a smart connectivity that is highly adaptive, ultrafast, and resilient for supporting secured interaction between humans and digital systems through the exploitation of novel terminals and mobile client nodes such as robots, mart connected cars, and drones [6]. 6G-BRAINS [7] proposes an AI-driven multiagent Deep Reinforcement Learning (DRL) architecture that performs resource allocation over and beyond mMTCs with new spectrum links including THz and optical wireless communications (OWC). 6G-BRAINS is to enhance the performance of 6G and beyond systems with regard to reliability, capacity, and latency for future industrial networks. The RISE-6G [8] project defines novel network architectures based on Reconfigurable Intelligent Surfaces (RISs) to create a new generation of dynamically programmable wireless propagation environments. The aim is to support dynamic adaptation to future stringent and highly varying 6G and beyond service requirements in terms of localization accuracy, electromagnetic field (EMF) emissions, and energy efficiency [8]. The key business drivers, applications, and use-cases pushing forward the development of 6G networks and beyond systems [4, 9] are summarized in Table 11.1. The evolution of communication services and applications toward 6G and beyond networks is indicated in Figure 11.1.

This include qualitative communications, HTC, High-Precision Communications (HPC), holographic teleport, Tactile Internet, tele-driving, cloud driving and integrated driving, Space-Terrestrial Integrated Network (STIN), Industrial IoT (IIoT) with fully cloudified Programmatic Logic Controllers (PLCs), automated traffic control systems, Intelligent Operation Network (ION), Light-Field 3D communications (LF3D), Network Computing Convergence (NCC), digital senses, and digital twin and holographic twin [5]. The 6G wireless network will provide extremely low-latency communications and support super-high-definition (SHD) and extremely high-definition (EHD) videos, that demand super-high throughput. 6G networks will also provide consistent multimedia streaming experiences even in emerging environments such as hyper-high-speed railway (HSR), natural disasters, and high-connected virtual shopping malls [4]. Table 11.1 shows a summary of different applications, usage scenarios, and 6G network characteristics. Figure 11.1 shows the evolution of communication services and applications toward 6G and beyond networks.

Table 11.1 A summary of applications, usage scenarios, and 6G networks characteristics.

Network category	Applications	Network characteristics	Usage scenarios
5G	VR/AR/360° Videos, UHD Videos, IoTs, Smart City/Industry/Factory/Home, Telemedicine, Wearable Devices.	Softwarization, Cloudization, Virtualization, Slicing.	enhanced Mobile BroadBand (eMBB); utrareliable and Low-Latency Communications (uRLLC), massive Machine-Type Communications (mMTC).
6G	Tactile/Haptic Internet, Holographic Verticals and Society, Full-Sensory Digital Sensing and Reality, Industrial Internet, Fully Automated Driving, Space Travel, Deep-Sea Sightseeing, Internet of Bio-Nano-Things, Robotic Automation, Holographic Media, Multisensory Holographic Teleportation, Real-time Remote Healthcare, Autonomous Cyber-Physical Systems, Intelligent Industrial and Robotic Automation, High-Performance Precision Agriculture, Space Connectivity, Smart Infrastructure and Environments.	Intelligentization, Cloudization, Softwarization, Virtualization, Slicing.	Extremely low-power communications (ELPC), Further-enhanced Mobile BroadBand (FeMBB), long-distance and high-mobility communications (LDHMC), ultramassive machine-type Communications (umMTC), extremely reliable and low-latency communications (ERLLC).

Table 11.2 Key performance indicators for 5G and 6G wireless networks.

Key performance indicators	5G	6G
Experienced data rate	0.1 Gb/s	1000 Gb/s
Peak data rate	201 Gb/s	⩾1Tb/s
Experienced spectral efficiency	3 × b/s/Hz	3 × b/s/Hz
Peak spectral efficiency	30 b/s/Hz	60 b/s/Hz
Connection density (Devices/km^2)	10^6 b/s/Hz	10^7
Area traffic capacity (Mbps/m^2)	10	1000
Maximum channel bandwidth (GHz)	1	100
End-to-end latency	1 ms	10–100μs
Mobility (km/h)	500	⩾1000
Delay Jitter	NA	10^-3
Network reliability (packet error rate)	10^-5	10^-9
Energy efficiency (Tb/J)	NA	1

11.1.1 Holographic and Future Media Communications

Holographic media [10] in 6G networks will need new form of communications over softwarized and virtualized systems such as holographic-type communications that are tolerant of quality degradation and characterized by very high throughput. Haptics and holograms will provide an immersive user experience even for multiple holographic streams. Hologram streaming in 6G networks will also support fast start-up and adapt to the changing network conditions in large bandwidth and low delay-supported automated networks [9]. The 6G network will have new packetization models that support high precision for time-based and qualitative services to manage throughputs. Holographic and full-sensory immersive experiences will lead the applications in 6G networks and in a variety of market verticals. New network-friendly media formats will be characterized by mechanisms to disaggregate volumetric data sets to object centric approaches with lots of metadata support. New holographic applications are expected to emerge in 6G networks and provide fully immersive AR/VR/XR experience with holograms [11]. 6G networks will meet the requirements of extremely high data rates in the order of Gbps or Tbps and stimulate all human senses (vision, hearing, smell, taste, touch, and balance) that will be important for conveying real-time user experience. New distributed HTC techniques proposed in [12] that perform adaptive signaling and frame buffering can be a starting point toward designing efficient algorithms for managing and improving the user's

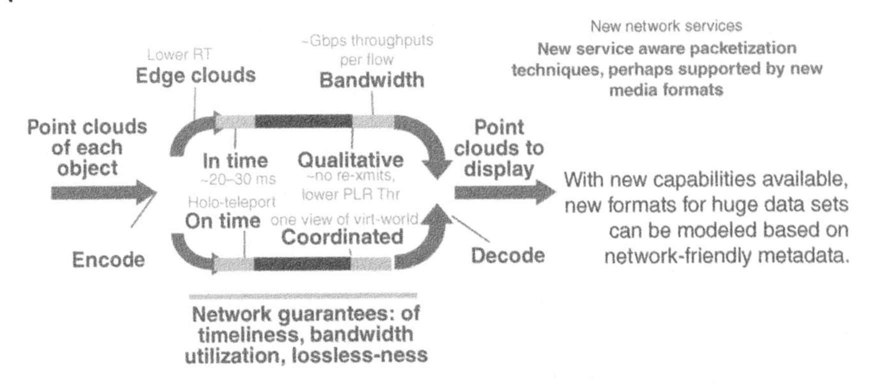

Figure 11.2 Holographic streaming for 6G and beyond networks.

QoE for teleportation streams. Figure 11.2 indicates the holographic streaming in 6G and beyond networks where new service packetization techniques, perhaps supported by new media formats will be dominant. New media formats for huge datasets will be modeled based on 6G network friendly metadata.

11.1.2 Human-Centric Services and 3D Volumetric Video Streaming

Human-centered service will be another class of services that needs intelligent, trusted, and inclusive quality models by considering physical factors from human physiology (brain cognition, body physiology, and gestures) [11]. The future of human-centric service in 6G and beyond networks will be supported by ML to provide meaningful one-to-one streaming experiences for users and allow them to receive great services via communication channels over Human-centric Intelligentized Multimedia Networking (HIMN). ML will enable utilizing communication, computing, and caching resources at different edge devices of 6G multimedia networks. The human's experience in HIMN in 6G systems will be put at the center of mobile video streaming services. Optimization of video quality will be done in real-time based on the human's content characteristics, context-awareness, and human's perceptions. New video codecs such as Versatile Video Coding (VVC) [13, 14] and other new video compression coding standards will play a central role in HIMN environments in 6G and beyond networks. Innovations in 5G and softwarized 6G networks provide new opportunities for human-centric video services for both consumers (e.g. cloud video gaming) and industries, particularly with respect to the multimedia sector.

The video evolution toward 8K and beyond in 6G networks will rely on the strict requirements such as experienced data rate and low E2E latency for video delivery.

However, with the development of new video coding standards (H.266)[1] for such services, a new set of QoPE metrics have to be defined and offered as mathematical function of traditional QoS and QoE metrics. The development of QoPE metrics models that learn human brain can be achieved using AI/ML and multiattribute utility theories from the operations research. A novel brain-aware learning and resource management approach proposed in [15] explicitly states that factors in the brain state of human users during resource allocation in a cellular network can be a starting point in developing the QoPE models in 6G networks.

As we move toward 6G and beyond networks, multimedia content is not only gaining higher video resolutions but also higher degrees of immersion. Volumetric video streaming is an emerging key technology to offer user interactions and immersive representation of 3D spaces and objects in 6G and beyond networks [16]. Volumetric videos provide viewers a 6DoF and 3D rendering, making them highly immersive, interactive, and expressive. 6DoF means three rotational dimensions (e.g. viewing direction in yaw, pitch, and roll) and three translational dimensions (e.g. viewpoint position in X, Y, and Z). Volumetric videos allow a viewer to freely change both the position in space and the orientation [17]. Streaming volumetric videos is highly bandwidth-demanding and requires lots of computational power because of their truly immersive nature.

Thanks to the experienced data rate and peak data rate (⩾1 Tb/s) offered by 6G networks which will play a significant part in delivering enough data fast enough for applications like volumetric video. Zhang et al. [18] propose a viewport prediction and blockage mitigation approaches that are efficient for streaming high-quality volumetric videos to multiple users. Authors provides a viewport-similarity opportunity that the multimedia research community can employ for optimizing effectively the network resource utilization using efficient multicast, and mmWave-aware, multi-user video rate adaptation. Qian et al. [17] introduce an efficient resource volumetric video streaming architecture that leverages edge computing over 5G/6G to reduce the computation burden on smartphone users while maintaining a high QoE. While the QoE metrics for regular videos have been well studied, but this remains to be an open problem for volumetric video streaming because the QoE of volumetric video streaming can be affected by factors such as viewing distance, visibility, point density, artifacts incurred by patches, motion-to-photon delay, and point density.

11.1.3 Potential Features and Technologies in 6G Networks

Several technologies are being explored from both academia and industry to support the requirements summarized in Table 11.2 and meet the vision of 6G

1 https://jvet.hhi.fraunhofer.de/

networks [3–5]. Figure 11.3 indicates key features and technologies for 6G and beyond networks. The design principles of 6G and beyond networks aim to use higher and unlicensed frequency bands to provide richer spectrum resources as well as multiplex more parallel data streams to achieve high spectral efficiency. In the realm of 6G communications and signal processing, AI-driven design and optimization will be applied to computer vision, cognitive radios, remote sensing, and network management. Advanced ML-based algorithms at the 6G network layer will be used for traffic clustering and adapt the network resources based on users' QoE demands as well as various scenarios. On the other hand, deep learning will be applied in 6G networks to optimize resource allocations at the physical layer for power distributions, modulations and coding, channel estimation, and multiuser detection.

These technologies include [3–5]: (i) Terahertz band communications, (ii) Large-scale multiuser Supermassive Multiple-Input Multiple-Output (SM-MIMO) systems, (iii) Holographic Beamforming (HBF) and RISs, (iv) Orbital Angular Momentum (OAM) multiplexing systems, (v) Laser and Visible-Light Communications (LVLC), (vi) blockchain-based spectrum sharing, (vii) the Internet of BioNanoThings, (viii) the Internet of NanoThings, (ix) Quantum communications and computing, (x) pervasive artificial intelligence, (xi) large-scale network automation and 3D network architectures, (xii) Ambient Backscatter Communications (Amb-BackCom) for energy savings, and (xiii) the Internet of Space Things (IoST) enabled by CubeSats and Unmanned Aerial Vehicle (UAV) [3]. The design principles of 6G and beyond networks aim to use higher and unlicensed frequency bands to provide richer spectrum resources as well as multiplex more parallel data streams to achieve high spectral efficiency. 6G networks will achieve nanoscale communications and interconnection and improve computing and energy efficiency. Figure 11.3 indicates key features and technologies for 6G and beyond networks. In the realm of 6G communications and signal processing, AI-driven design and optimization will be applied to computer vision, cognitive radios, remote sensing, and network management. Advanced ML-based algorithms at the 6G network layer will be used for traffic clustering and adapt the network resources based on users' QoE demands as well as various scenarios. On the other hand, deep learning will be applied in 6G networks to optimize resource allocations at the physical layer for power distributions, modulations and coding, channel estimation, and multiuser detection.

Bolstered by the use of SDN and NFV, the IoST is envisaged to achieve a global connectivity at low costs and signifies a cyber-physical system that spans the air, ground, and in-space backhauling as well as holistic data integration in 6G and beyond networks. The space segment of IoST consists of UAV, CubeSats, and near-Earth sensing devices, while the ground segment is formed by ground station, on-Earth sensing devices, and customer premises [3]. Different from

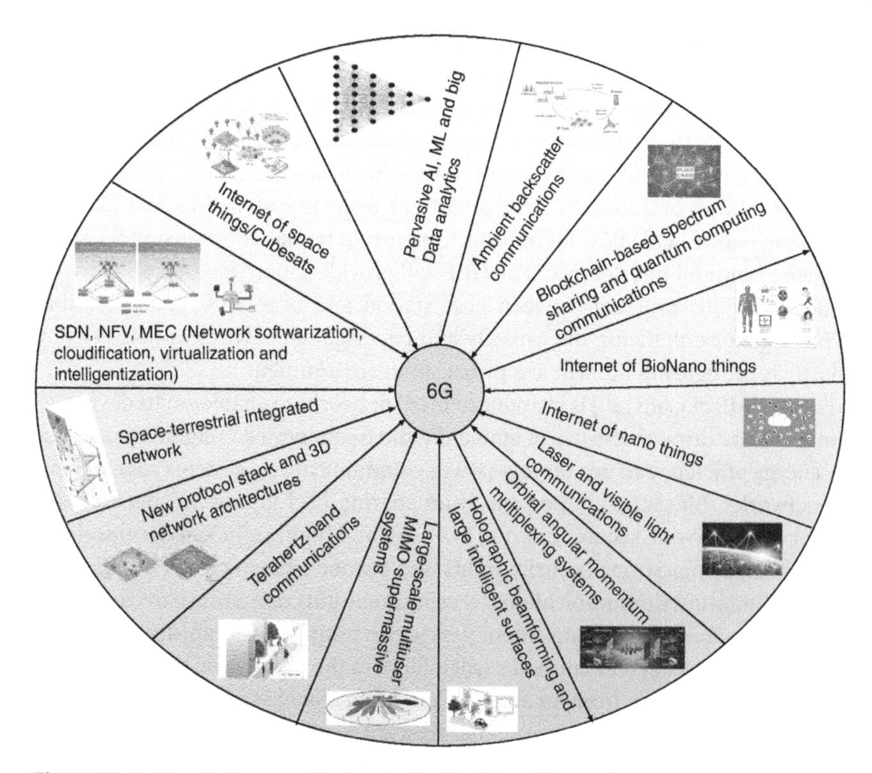

Figure 11.3 Key features and technologies for 6G and beyond networks. Source: (i) Sdecoret/Adobe Stock; (ii) Iconimage/Adobe Stock.

mmWave communications [19] in 5G networks, THz communications in 6G and beyond promise to enable ultrahigh bandwidth communication paradigms by providing Tbps links for various services and applications including high-definition videoconferencing among mobile devices in small cells environment. Large-scale SM-MIMO with huge multiplexing gain and beamforming capabilities will allow data/information collected from smart sensors to be transmitted with lower latency while offering higher data rates and reliable connectivity to the end user's devices in 6G and beyond networks.

The SM-MIMO will significantly improve energy efficiency and achieve superhigh spectrum efficiency by using spatial multiplexing techniques that transmit hundreds of parallel data streams on the same frequency channel [5]. Traditional wireless communications using electromagnetic wave signals will not be enough to provide high-speed data transmissions in 6G networks that will encompass underwater networks and space/air networks with terrestrial networks. It is worth noting that laser communications in 6G networks and beyond will provide ultrahigh bandwidth and achieve high-speed data transmission using laser beams, which are suitable for environments such as free space and under water. HBF is a new dynamic beamforming approach which employs a Software Defined Antenna (SDA) which will enable wireless service providers to reuse continuously abundant spectrum with higher intensity signals delivered to both stationary and mobile users using the lowest cost, size, weight, and power (C-SWaP) architecture. Unlike current cellular systems, HBF in 6G and beyond networks will allow for multiple concurrent transmissions using the same frequency without interference. The HBF will provide a more focused 6G network communications protocol between base station and user. RISs is a promising technology for enhancing the capacity and coverage of 6G wireless networks by smartly reconfiguring the wireless propagation environment.

The Amb-BackCom [20] is introduced in 6G networks to enable smart devices to communicate through the use of ambient radio frequency (RF) signals to address the energy efficiency issues for low-power communications systems such as sensor networks. Blockchain-based spectrum sharing [21] is a promising technology for 6G to provide secure, smarter, low-cost, and highly efficient decentralized spectrum sharing. Strong security in 6G and beyond networks will be provided through Quantum Communications by using quantum computing based on quantum superposition and entanglement. It is worth noting that quantum communications in 6G and beyond networks will adhere to the no-cloning theorem where no copies can be made from an arbitrary quantum state [3]. Unlike conventional networks that are based on the store-and-forward paradigm, 6G and beyond-based quantum networks will apply the quantum teleportation process in transmitting unknown quantum states between remote quantum devices [22, 23]. We provide the description of self-driving 3D network architecture, pervasive artificial intelligence, network management, automation, and orchestration [24].

11.2 6G Innovative Network Architectures: From Network Softwarization to Intelligentization

Figure 11.4 depicts the primary architectural advances that 6G will bring forth. We envision the following concepts being introduced and/or used beyond network softwarization aspects to intelligentization in 6G networks. We provide aspects of 6G innovations that leverage the network management, automation and orchestration, pervasive AI, ML, and big data analytics, new protocol stack, and 3D network architectures.

11.2.1 Network Management, Automation, and Orchestration

Network automation using AI/ML will build upon the improvements offered by SDN and NFV and other technologies discussed in Chapter 11 to speed up the

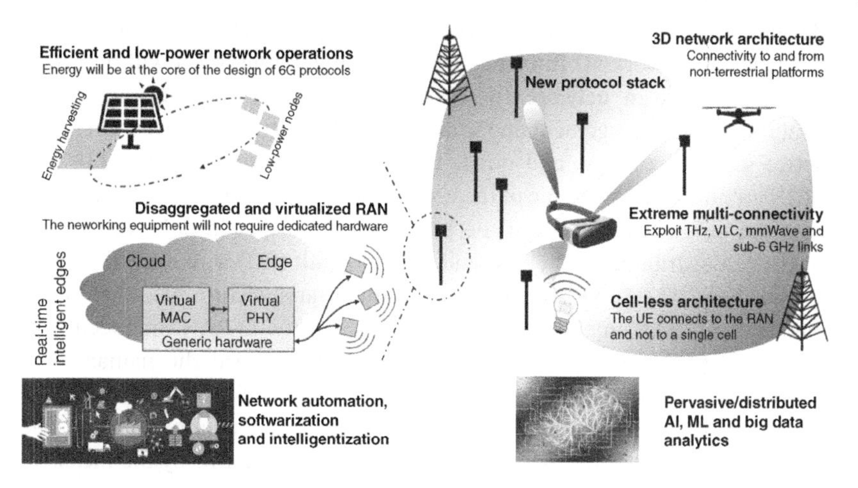

Figure 11.4 Architectural innovations and mechanisms introduced in 6G networks. Source: Geralt/Pixabay.

delivery of 6G network media services while guaranteeing experienced QoE to the users. ML will provide the needed functionality to guarantee automation of future 6G wireless communication networks [25]. The management and orchestration of 6G systems using ML-based automation will enable real-time analysis prediction and automated zero-touch operation, management, and control in 6G and beyond networks [26]. The intelligence embedded in different communication nodes will rely on the availability of data timely streamed from wireless devices, especially in extreme services and applications such as XR, real-time video streaming, and holographic media. Various categories of ML including supervised learning, unsupervised learning, and supervised learning will enable several advanced services and QoS/QoE functionalities related to traffic prediction, and classification as well as predicting the QoE resource requirements associated with different network slice based on the anticipated traffic load. AI/ML will continue to be an indispensable tool in the network management domain, network security, and the end user's QoE optimization in 6G and beyond networks [11].

The introduction of Management Data Analytics Service (MDAS) and the Network Data Analytics Function (NWDAF) within 3GPP Releases 15 and 16 which together form an important part of the 5G Service-Based Architecture (SBA). The 3GPP defines a SBA, whereby the common data repositories and control plane functionality of a 5G network are offered and delivered by way of a set of interconnected NFs, each with authorization to access each other's services. The SBA in the 5G/6G network and beyond will allow on demand deployment of each service while being updated independently with minimal impact to other services [3]. It will also allow automation and agile operational processes, vendor independence, and reduction in services delivery and enhanced operational efficiencies of network 6G functions. Network automation in 6G will transition from operator-driven network to self-driving networks and allow error-free operation, elastic management and utilization of resources, proactive rather than reactive service handling, and quick responses to security incidents [3]. The O-RAN architecture [27] is one of the evolutionary steps from 5G to 6G networks where the different functions of the base station are split into the centralized unit (CU), a distributed unit (DU), and a remote unit (RU) with open interfaces between them. The 3GPP also proposes a similar architecture like the O-RAN approach. The RAN intelligent controller (RIC) in the O-RAN architecture is extracted from the processing units and allows it to reach the management interfaces, like radio resource management (RRM) or self-organizing networks (SON) functions, which control the radio resources and network operation. The intelligence in the O-RAN concept lies is positioned in the RIC by the means of AI models for radio network automation.

11.2.2 Pervasive AI, ML, and Big Data Analytics

AI/ML will be a key component of 6G system functioning to reduce or eliminate human intervention. It is projected that the 6G system will generate, transmit, store, and use an unprecedented amount of data as a result. Part of these data will be user-related data. It will be essential to safeguard the privacy, authenticity, and provenance of these data if 6G is to be trusted and, more critically, if 6G is to function as intended (e.g. any malicious data corruption could have detrimental effects on the operation of 6G). The development of 6G and beyond networks with a new convergence of Communication, Computing, and Caching (3C) requires intelligent approaches that can keep the KPI within the required thresholds and meet the changing network conditions [28]. Along with cutting-edge technologies (SDN, NFV, and MEC), AI/ML will play a central role in the control, management, orchestration, and full automation of 6G networks. AI/ML can be applied at different layers of 6G networks. ML strategies can be used at the network layer to perform network traffic clustering and adapt network resources based on the user's demands. Resource allocation algorithms can be optimized using ML in the 6G radio protocol stack [29].

AI-empowered 6G will support several features including self-configuration and optimization, sophisticated learning, aggregation, opportunistic set-up, knowledge discovery, and QoE-based context awareness to users [11]. AI-assisted brokering mechanism for RAN slicing and AI-empowered slice admission control, slice scheduling, handover management, and mobility management will be prevalent in the era of 6G networks [26]. AI in RAN will help to optimize the network resources and enable real-time conversations among the 6G network entities. In terms of operation and management, AI-based 6G networks will enable intelligent measurement and monitoring as well as enhanced security. New AI-empowered data plane techniques have to be developed to support HTC and ITS in 6G networks that will be able to evolve and adapt dynamically to changing network conditions as well as user's service quality demands while transmitting holographic-payloads [11].

11.2.3 New Protocol Stack and 3D Network Architectures

The 6G networks and beyond architecture will complement and integrate both terrestrial and nonterrestrial platforms such as low-orbit satellites, balloons, and drones. While previous generations including 5G networks were designed to support services in two-dimensions (2D), 6G and beyond networks require a 3D architecture to provide ubiquitous 3D coverage, seamless and extremely service connectivity, pervasive connectivity, unmanned mobility, holographic telepresence support. Again, the concept of 5G network slicing should be extended in 6G networks such that it can be applied across both terrestrial and nonterrestrial nodes. Several projects (e.g. H2020 VIrtualized hybrid satellite-TerrestriAl systems for resilient and fLexible future networks and SANSA) have been initiated to develop 3D-based 6G and beyond network architecture. The VITAL project develops a hybrid 6G architecture that integrates terrestrial and satellite networks by bringing the NFV into the satellite domain and by enabling SDN-based, federated resources management in hybrid SatCom-terrestrial networks.

The VITAL project aims at optimizing the network communications resources usage, improving the coverage, and providing better network resilience. The H2020 SANSA project provides a terrestrial-satellite backhaul architecture that supports capabilities of future terrestrial 6G wireless networks to reconfigure automatically based on the changing traffic demands. The SANSA architecture will also support a shared spectrum between satellite and terrestrial segments and a seamless integration of the satellite segment into terrestrial backhaul networks. The TCP/IP protocol stack for many years has enabled the evolution of connected computing by providing network connectivity and transmission of applications and services for end users. Standard bodies such as IETF and 3GPP have been the driving force behind the incremental improvements of the TCP/IP protocol suite by solving problems including the introduction of IPv6 and the introduction of HTTPS to enhance security and privacy. However, this has been done in a segmented manner through the addition of new protocols by focusing on a specific service requirement or protocol layer while keeping its core structure unchanged. Whenever there is a new requirement or a new application to be supported, a new protocol is designed, implemented, and deployed. The communication demands of emerging new applications and services in future

6G and beyond networks become increasingly difficult to address them using existing protocols and internetworking technology. Many infrastructures running future communication protocols will be based on virtualized and softwarized architectures. This calls for new data-driven network and protocol architectures and removes the anchor drag of historic suboptimal IP protocol stacks which are single-dimensional, built for interconnectivity of topology-centric fixed devices.

11.3 6G Standardization Activities

The academia and industry have started to establish the service requirements for the coming decade. For example, the ITU-T with the Focus Group on Network 2030 [30] have been developing the next steps from the current 5G networks. In addition, beyond-5G solutions are beginning to be created based on research and innovation projects across many locations including the 6G flagship, [2] HEXA-X.[3] These early experiences, together with other recent industrial projects like NGMN [31], are anticipated to be very helpful to standardization organizations supporting the creation of cutting-edge 6G systems. Surprisingly, ITU-R has just established a vision group on IMT toward 2030 and beyond 6G, with the goal of finishing its work by the end of 2023. Figure 11.5 indicates the overall roadmap for 6G development toward 2030 networks.

From the perspective of 3GPP standards development, there are still several features and capabilities from current 5G solutions that need full definition; these are anticipated to be finished in the upcoming 3GPP Release 18, which is targeted for release at the end of 2023. Next releases are anticipated to be focused on the evolution of 5G by the middle of the decade, or 2025, followed by an analysis of 6G and then the correct 6G standard. The general development schedule for 6G is shown in the following diagram [32].

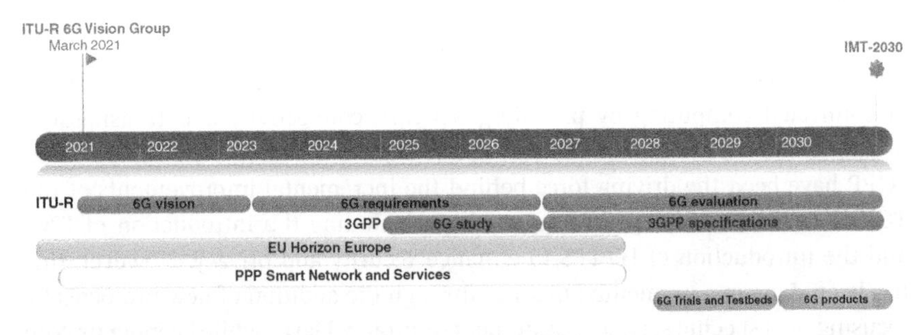

Figure 11.5 The overall roadmap for 6G development toward 2030 networks.

2 https://www.6gflagship.com
3 https://hexa-x.eu

11.4 Conclusion

This chapter builds on the achievements of 5G networks to provide the first roadmap toward 6G and beyond networks in terms of requirements, use-case scenarios, and service classes (e.g. holographic and future media communications, human-centric services and 3D volumetric video streaming, and new video compression standards). The chapter presents potential technologies that will be dominant through 2025–2030 in shaping the vision of 6G and beyond networks and multimedia streaming services delivery. Furthermore, the chapter provides various features and 6G innovative network architectures that leverage intelligentization including pervasive AI/ML and big data analytics. It also provides 6G standardization activities and challenges as well as research directions in 6G and beyond networks. The comprehensive goal of this chapter is to encourage researchers from the industry and academia to work together toward the realization of softwarized 6G and beyond networks while tackling the critical research challenges regarding management of emerging 3D multimedia services and applications.

Bibliography

1. Andrews, J.G., Buzzi, S., Choi, W. et al. (2014). What will 5G be? *IEEE Journal on Selected Areas in Communications* 32 (6): 1065–1082.
2. Agiwal, M., Roy, A., and Saxena, N. (2016). Next generation 5G wireless networks: a comprehensive survey. *IEEE Communication Surveys and Tutorials* 18 (3): 1617–1655.
3. Akyildiz, I.F., Kak, A., and Nie, S. (2020). 6G and beyond: the future of wireless communications systems. *IEEE Access* 8: 13399–134030.
4. Saad, W., Bennis, M., and Chen, M. (2020). A vision of 6G wireless systems: applications, trends, technologies, and open research problems. *IEEE Network* 34 (3): 134–142.
5. Zhang, Z., Xiao, Y., Ma, Z. et al. (2019). 6G wireless networks: vision, requirements, architecture, and key technologies. *IEEE Vehicular Technology Magazine* 14 (3): 28–41.
6. DEDICAT 6G: Dynamic coverage Extension and Distributed Intelligence for human Centric Applications with assured security, privacy, and Trust: from 5G to 6G. https://5g-ppp.eu/dedicat-6g// (accessed 22 May 2021).
7. 6G BRAINS: Bringing Reinforcement learning Into Radio Light Network for Massive Connections. https://5g-ppp.eu/6g-brains// (accessed 22 May 2021).
8. RISE-6G: Reconfigurable Intelligent Sustainable Environments for 6G Wireless Networks. https://5g-ppp.eu/rise-6g// (accessed 22 May 2021).

9 Network 2030: A Blueprint of Technology, Applications and Market Drivers Toward the Year 2030 and Beyond. https://www.itu.int/en/ITU-T/focusgroups/net2030/Documents/White.Paper.pdf (accessed 13 August 2022).

10 Clemm, A., Vega M. T., Ravuri H. K., et al. (2015). Toward truly immersive holographic-type communication: challenges and solutions. *IEEE Communications Magazine* 58 (1): 93–99.

11 Letaief, K.B., Chen, W., Shi, Y. et al. (2019). The roadmap to 6G: AI empowered wireless networks. *IEEE Communications Magazine* 57 (8): 84–90.

12 Selinis, I., Wang, N., Da, B. et al. (2020). On the internet-scale streaming of holographic-type content with assured user quality of experiences. *IFIP Networking Conference (Networking)*, July 2020.

13 New 'Versatile Video Coding' standard to enable next-generation video compression. https://www.itu.int/en/mediacentre/Pages/pr13-2020-New-Versatile-Video-coding-standard-video-compression.aspx (accessed 13 August 2022).

14 Huang, Y.-W., Hsu, C.-W., Che, C.-Y. et al. (2020). A VVC proposal with quaternary tree plus binary-ternary tree coding block structure and advanced coding techniques. *EEE Transactions on Circuits and Systems for Video Technology* 30 (5): 1311–1325.

15 Kasgari, A.T.Z., Saad, W., and Debbah, M. (2019). Human-in-the-loop wireless communications: machine learning and brain-aware resource management. *IEEE Transactions on Communications* 43 (11): 7727–7743.

16 Gül S., Podborski, D., Buchholz, T. et al. (2020). Low-latency cloud-based volumetric video streaming using head motion prediction. *Proceedings of the 30th ACM Workshop on Network and Operating Systems Support for Digital Audio and Video*, pp. 27–33, June 2020.

17 Qian, F., Han, B., Pair, J., and Gopalakrishnan, V. (2019). Toward practical volumetric video streaming on commodity smartphones. *Proceedings of the 20th International Workshop on Mobile Computing Systems and Applications*, pp. 135–140, February 2019.

18 Zhang, A. et al. (2021). Mobile volumetric video streaming enhanced by super resolution. *Proceedings of the 20th ACM Workshop on Hot Topics in Networks*, pp. 16–22, November 2021.

19 Niu, Y. et al. (2017). A survey of millimeter wave communications (mmWave) for 5G: opportunities and challenges. *Wireless Networks* 21: 2657–2676.

20 Van Huynh, N. et al. (2018). Ambient backscatter communications: a contemporary survey. *IEEE Communication Surveys and Tutorials* 20 (4): 2889–2922.

21 Kotobi, K. et al. (2018). Secure blockchains for dynamic spectrum access: a decentralized database in moving cognitive radio networks enhances security and user access. *IEEE Vehicular Technology Magazine* 13 (1): 32–39.

22 Caleffi, M. et al. (2018). Quantum Internet: from communication to distributed computing. *Proceedings of the 5th ACM International Conference Nanoscale Computing and Communication (NANOCOM)*, pp. 1–4, September 2018.

23 Nawaz, S.J. et al. (2019). Quantum machine learning for 6G communication networks: state-of-the-art and vision for the future. *IEEE Access* 7: 46317–46350.

24 Akhtar, M.W., Hassan, S.A., Ghaffar, R. et al. (2020). The shift to 6G communications: vision and requirements. *Human-centric Computing and Information Sciences* 10 (53): 1–27.

25 Zhang, C., Patras, P., and Haddadi, H. (2019). Deep learning in mobile and wireless networking: a survey. *IEEE Communication Surveys and Tutorials* 21 (3): 2224–2287.

26 Yang, H., Alphones, A., Xiong, Z. et al. (2020). Artificial-intelligence-enabled intelligent 6G networks. *IEEE Network* 34 (6): 272–280.

27 O-RAN Architecture Overview. https://docs.o-ran-sc.org/en/latest/architecture/architecture.html (accessed 22 June 2021).

28 Benzaid, C. and Taleb, T. (2020). AI-driven zero touch network and service management in 5G and beyond: challenges and research directions. *IEEE Network* 34 (2): 186–194.

29 Bega, D. et al. (2020). AI-based autonomous control, management, and orchestration in 5G: from standards to algorithms. *IEEE Network* 34 (6): 14–20.

30 FG-2030. Focus Group on Technologies for Network 2030. https://www.itu.int/en/ITU-T/focusgroups/net2030/Pages/default.aspx (accessed 06 May 2022).

31 NGMN 6G Drivers and Vision. https://www.ngmn.org/work-programme/ngmn-6g-drivers-and-vision.html (accessed 13 August 2022).

32 European Vision for the 6G Network Ecosystem. https://5g-ppp.eu/wp-content/uploads/2021/06/WhitePaper-6G-Europe.pdf (accessed 13 August 2022).

12

Multimedia Streaming Services Delivery in 2030 and Beyond Networks

This chapter provides the multimedia streaming services delivery aspects in the 2030 and beyond networks. The chapter focuses on new capabilities, features of the communications networks and services required from 2030, and beyond networks. We present an OTT/CDN-ISP collaboration for QoS-aware multi-CDN adaptive video streaming in future 2030 and beyond networks.

12.1 The Future of the Video Streaming Industry: Market Growth and Trends Toward 2030

The global video streaming market size was valued at USD 59.14 billion in 2021 and is expected to reach USD 223.98 billion by 2030 [1]. The market is to expand at a CAGR of 21.3% from 2022 to 2030. The technological advancements such as the implementation of blockchain technology in video streaming and the use of artificial intelligence for improving video quality are defiantly to boost the demand for the market over the 2030 forecast period. Furthermore, the growing adoption of cloud-based video streaming solutions for increasing the reach of video content is directly influencing the growth. This trend is observed in numerous parts of the North American and Asia Pacific regions. The reasons attributing to the growth of the market are rapid digitalization, increasing use of mobiles and tablets, and the growing popularity of online video streaming.

The factors such as the rising demand for on-demand video and extensive growth of online video are the key growth drivers for the market toward 2030 communication networks. Also, the increasing demand for high-speed Internet connectivity acts as an advantage for the market to grow over the forecast period. The growing acceptance of smartphones in combination with an extensive range of high-speed Internet technologies such as 5G [2], and 6G [3] has substantially led to the increasing use of video streaming. Also, the growing demand for devices that can support digital media is helping consumers access media content

Multimedia Streaming in SDN/NFV and 5G Networks: Machine Learning for Managing Big Data Streaming, First Edition. Alcardo Barakabitze and Andrew Hines.

anywhere across the world [4]. While heading toward multimedia services delivery in 2030 networks, content creators and broadcasters cannot continue to rely on their current market position. They must be willing to work together and form alliances, even with direct rivals, in order to protect their video quality business models and potential revenue sources. The danger posed by digital platform providers such as Netflix, Amazon, Apple, or Google can be countered by cooperative multimedia streaming delivery, QoE-driven joint distribution approaches, and even shared platforms that leverages intelligent softwarized and virtualized networks [5].

12.2 Future 2030 Communication Network and Computing Scenarios

A key component of Society 5.0, Knowledged Human Bond Communications Beyond 2050 (Knowledge Home) [6] and the integration of Communication, Navigation, Sensing, and Services (CONASENSE) [7] are all examples of the human-centric aspects of 2030 networks. Another feature of 2030 is intelligence, which calls for the core and the edge of the innovative network to have understanding of how to increase traffic. Another characteristic of 2030 communication and beyond networks is intelligence, where the edge and core of the innovative network will need to be intelligent about how to increase traffic [8]. ML and AI are prerequisites for communication networks of 2030 and beyond. To enable an intelligent network handling big data in near real-time and optimize network traffic prioritization based on network slicing and other particular KPIs, AI, and ML will be essential. Therefore, the capacity of learning and adapting the Experience-Level Agreements (ELAs) [9] for different QoE for multimedia services will be of utmost. In order to meet the demands of big data, Quantum Computing (QC) is a technology that will be integrated into the edge of 6G networks. If science and technology could develop QC to the point where it could be used on a big basis over the next 10 years, it would be beneficial [3]. Figure 12.1 indicates future computing scenarios for 2030 communication and beyond networks.

The real and digital worlds will effortlessly mix in 2030. Perceptual and emotional interactions between people and machines will occur. The physical world will be simulated, improved, and recreated by new computing paradigms. Hyper-realistic experiences will push computing to the edge and call for multidimensional collaborative computing between the cloud and edge, between edge and edge, and between the virtual and actual worlds [4]. Perceptual intelligence will give way to cognitive intelligence as AI develops its creative abilities. Everything will become smarter and more inclusive as a result [3]. There will be 200 billion connections by 2030. Information regarding the physical world, such as

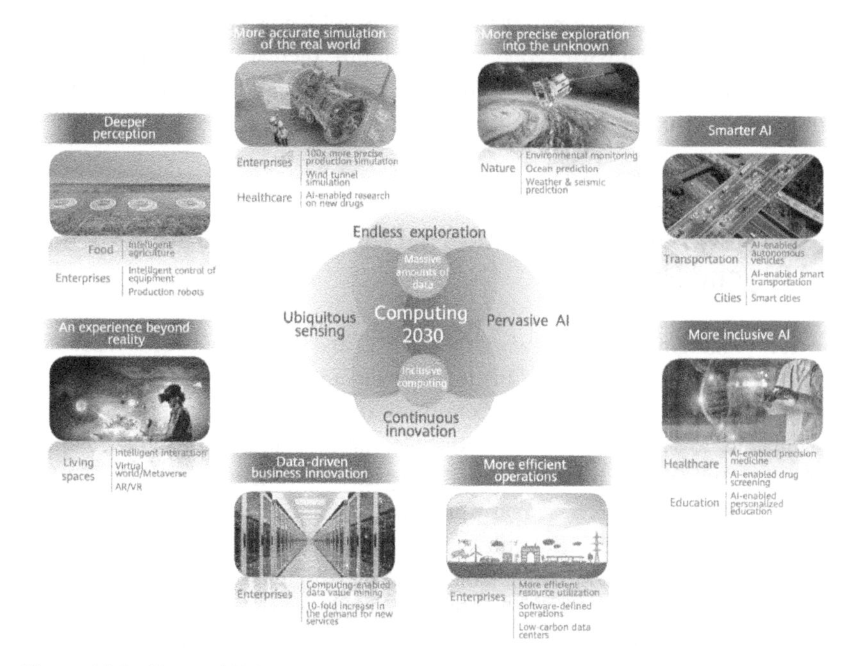

Figure 12.1 Future 2030 communication networks and computing scenarios. Source: (i) Kosssmosss/Adobe Stock; (ii) Sergey Nivens/Adobe Stock; (iii) Gorodenkoff/Adobe Stock; (iv) Armyagov/Adobe Stock; (v) Metamorworks/Adobe Stock; (vi) Ipopba/Adobe Stock; (vii) Sdecoret/Adobe Stock; (viii) Elnur/Adobe Stock.

temperature, pressure, speed, light, humidity, and chemical concentration, will be gathered by hundreds of trillions of sensors. Deeper perceptual abilities will be needed to convert this fundamental data into sensory information so that robots may see, hear, taste, smell, and feel things. The process of computation for creating sensory information must be finished at the edge due to problems with data amount and latency. Therefore, the edge will need to be able to handle data intelligently, which would entail modeling information processing in the human brain. The edge, where around 80% of data will be handled, will perform a significant amount of perceptual computing in the future.

A new core element called Quantum Machine Learning (QML) will be created by combining QC and machine learning [8]. This calls for the use of QML to accelerate the learning of all transactions occurring at the 6G Network's edge in order to automatically change the network's resources. Because 6G will be a decentralized network with a need for faster processing capability to manage all the enormous data traffic anticipated to flow in the network, QML is being used. The evolution of 8K will be 14K/16K video streaming, which requires will require 10 times the amount of bitrate of the current video format. For video streaming different from 3D video format, the bitrate is even higher. The processing power available in QML will easily cope with the future increase of video traffic and heterogeneous data. QML's role in the 2030 communication network will be to optimize the video streaming traffic for the 2030 applications and services [8].

12.3 New Paradigms of Internetworking for 2030 and Beyond Networks

The new paradigms of internetworking for 2030 and beyond networks discussed in this section as indicated in Figure 12.2 include the following: Qualitative Communications, High-Precision Communications, and Holographic Teleport. The capacity to send and interact with holographic data from distant locations across a network will be made possible by holographic-type communication (HTC). Such applications have many practical uses and are far from being a gimmick [1]. To promote real-time interactions with local participants, holographic telepresence, for instance will enable remote participants to be projected as holograms into a meeting room. Similar to this, training and educational software can let users engage dynamically with holographic objects that are incredibly lifelike for teaching reasons [10]. When needed, information must always be instantly accessible within the constraints of time. The amount of data being produced these days was briefly mentioned in earlier parts. Networked storage can be utilized to simulate situation models (almost like digital twins of events) with this much data, and when such situation models turn catastrophic, instant real-time judgments can be forcedly conveyed to different system entities. Thus,

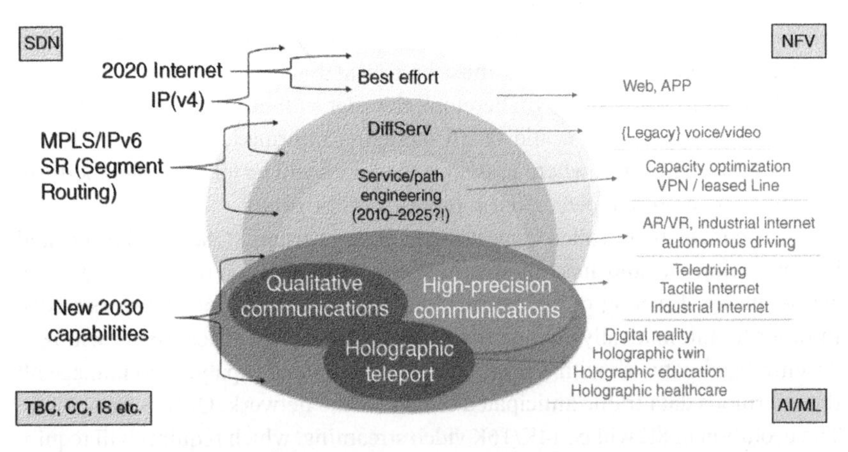

Figure 12.2 New capabilities and services required for 2030 and beyond networks.

by the year 2030, networks should be able to distinguish between the urgency and purpose of this data and transmit it accordingly [11].

12.3.1 Next-Generation Human–Machine Interaction Network: A Human-centric Hyper-real Experience

In the 2030 era, it will be feasible to flip the new paradigms of inter-networking and have computers/machines adapt to the demands of their human users with sufficiently high levels of intelligence. Intelligent machines will be able to comprehend natural language, gestures, eye movement, and even read human brain waves. This will allow for more intuitive integration between the virtual and physical worlds and bring a hyper-real sensory experience to human–machine interaction. Examples of such machines include smart screens, smart home appliances, intelligent vehicles, and smart exoskeletons. Figure 12.3 indicates the hyper-real human–machine interaction experience. Communications networks will need to change over the course of the next 10 years to enable cutting-edge human–machine interface techniques including XR, naked-eye 3D displays, digital touch, and digital odor. Communications networks will need to change over the course of the next 10 years to enable cutting-edge human–machine interface techniques including XR, naked-eye 3D displays, digital touch, and digital odor [12].

An sophisticated kind of augmented reality (AR) called mixed reality (MR) incorporates virtual components into real-world situations. A general phrase used to describe both real and virtual mixed worlds and human–machine interactions produced by computer technology and wearables is called "eXtended Reality," which encompasses VR, AR, and MR. XR is seen as the next significant platform for interpersonal interactions because of its three-dimensional surroundings, intuitive interactions, spatial computing, and other aspects that make it unique from current Internet gadgets [4]. The technical architecture of XR is divided into five components by the China Academy of Information and Communications Technology (CAICT): near-eye display, perception and interaction, network transmission, rendering processing, and content development. Today, a typical XR experience involves 2K monocular resolution, 100°–120° FOV, 100 Mbit/s bitrate, and 20 ms motion-to-photon (MTP) latency. If all content is rendered in the cloud, 20 ms of MTP latency is the threshold above which users start to report feelings of dizziness [13].

By 2030, XR will be at the point of complete immersion thanks to 8K monocular resolution, 200° FOV, and gigabit-level bitrate capability [12]. As long as rendering is still done entirely in the cloud, MTP latency must be kept under 5 ms. The delay will be directly tied to the sorts of content if technology is created to facilitate the local display of environment-related content that could easily make viewers dizzy. Twenty milliseconds of MTP latency is sufficient for content that just needs

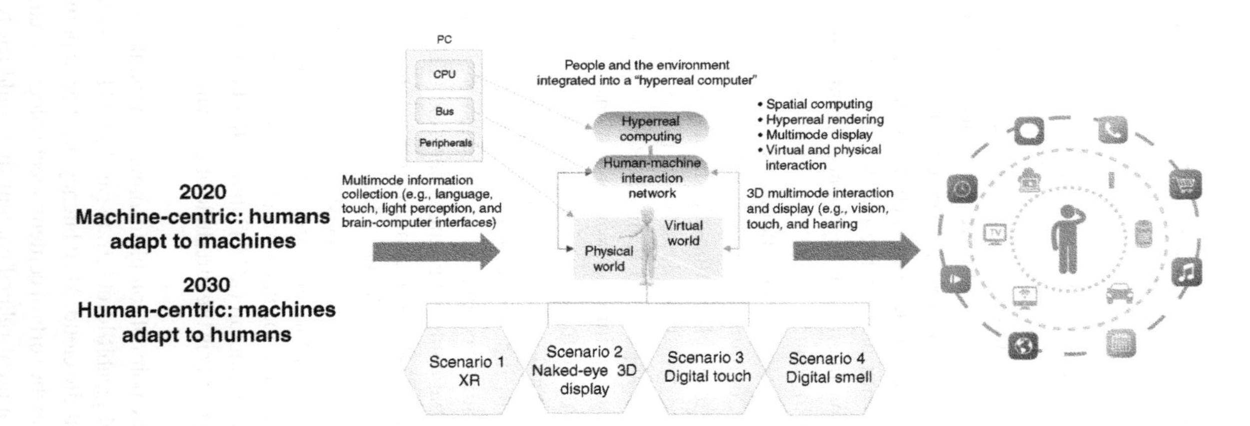

Figure 12.3 Hyper-real human–machine interaction experience.

minimal interaction (like a streamed video). Less than 5 ms MTP latency will be required for materials like games that demands intense engagement. Therefore, to support the development of XR services over the next 10 years, networks must have bandwidth of at least 1 Gbit/s and latency of less than either 5 or 20 ms, depending on the scenario [3].

12.3.2 Multimedia Streaming Blockchain-enabled Resource Management and Sharing for 2030 and Beyond Networks

Most of video streaming platforms today are still based on the client-server model of the content delivery networks which have some major drawbacks including Flash crowd where a lot of streaming servers cannot feed more than a hundred streaming sessions at the same time; single point of failure and bandwidth cost. Blockchain technology works as a peer-to-peer network, eliminating any central authority. P2P network provides features that do not focus on one single server but work by distributing the media content. Video content curators benefit from the decentralized mechanism of blockchain because they directly publish and deliver the video content to their audience without any intermediary intervention or central server.

Because it uses a P2P framework, video streaming is significantly better when done on a decentralized network. A P2P network increases the likelihood that one stream will divide and become dispersed. This occurs as a result of P2P having a higher video streaming replication rate than client–server CDN structures. Both creators and audiences can join the blockchain network without the need for a middle controller. The network is used by the content curators to independently broadcast their content, which is then viewed by the public after being approved by the other nodes. Blockchain-based video streaming enables content creators to gain direct access to the revenue produced by the blockchain network. The public only pays for the content they are most interested in viewing; therefore, it allows the content curators to offer videos based on audience demand. The blockchain can be used by content producers to safely encrypt their video storage [8].

Blockchain enables content curators to receive payments from users in the form of cryptocurrency tokens. When viewers pay membership fees to their preferred streams, they receive payment. Additionally, if viewers choose to rent their extra disk space and network bandwidth for the purpose of storing and broadcasting the movies, they can also profit handsomely from doing so [14]. The various ways in which the blockchain will support and reinvents online video streaming in 2030 and beyond networks are as follows: Ownership and accessibility, blockchain-based CDN, smart contracts, content licensing, and effective monetization. With the help of a blockchain-based management system, the content creators can use the collective unused space by converting the space into P2P cloud storage and data delivery system using blockchain-based CDN. With content licensing and effective monetization, virtual content creators will be aware of the implications of intellectual property laws and the convenience with which their work may be stolen or copied.

In terms of ownership and accessibility, with the help of blockchain, the issue of ownership can be easily addressed at a protocol level. Blockchain helps the creators and the audience encrypt their videos and store those videos permanently. Blockchain will be able to broadcast the video content in an autonomous manner with the help of various relay nodes, thereby maintaining transparency and integrity of ownership in a decentralized way [15]. Blockchain will also maintain ownership of the video streamers in the form of nonfungible tokens (NFTs). People who broadcast their video content will also have the access to trade it on the marketplaces in exchange for cryptocurrencies in 2030 networks. They will also secure their video content ownership via tokenizing their content on the platform.

Figure 12.4 shows an illustration of a blockchain-enabled multimedia streaming resource management framework in 2030 networks. The programmable blockchain feature, which is typically referred to as a "smart contract," is what the automated blockchain-enabled resource management is dependent on. The content of the contract is open to the public and the parties to the agreement, making it publicly tracable. The blockchain has opened up a new market for trading

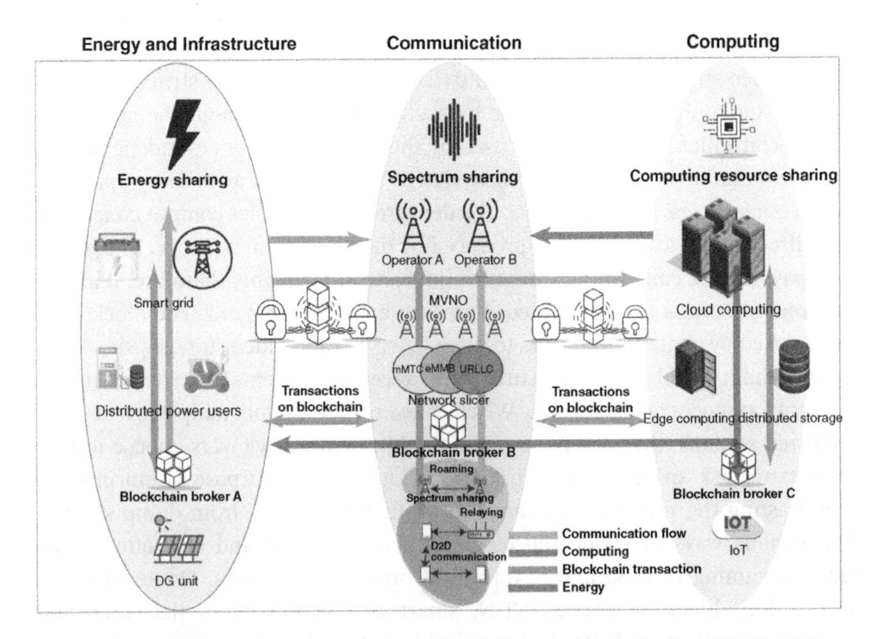

Figure 12.4 Blockchain-enabled multimedia streaming resource management framework in 2030 networks.

resources for fixed assets such as computer hardware and licensed spectrum. Tradable spectrum and computing resources are integrated into the resource pool in the blockchain-enabled 6G resource management framework shown in Figure 11.4. Spectrum is dynamically allocated, network slices are managed, and the hardware is virtualized to facilitate blockchain-enabled resource management [14]. Smart contracts are executed using the virtual machine paradigm, where the code is run by a node on the virtual stack, and the outcomes are recorded as transactions on the chain. The contract has a high degree of immutability against violations of the contract and misrepresentations thanks to its temper-proof capability and totally automatic process [14].

12.4 A General QoE Provisioning Ecosystem for 2030 and Beyond Networks

The QoE will be the main driving performance metric in terms of multimedia services delivery in 5G softwarized networks. This is so because of the increasing number of multimedia services/applications and the demand of high QoE from the end users. To achieve this, different key stakeholders have to be involved in the general QoE provisioning ecosystem in future softwarized 5G and beyond networks. To meet users' demand, we introduce the QoE provisioning ecosystem in 5G networks shown in Figure 12.5. The 5G communication ecosystem that involves all stakeholders in the QoE delivery chain in future softwarized 5G and beyond networks is shown in Figure 12.5. We envision that video content providers should prepare the actual video content as digital items by considering essential

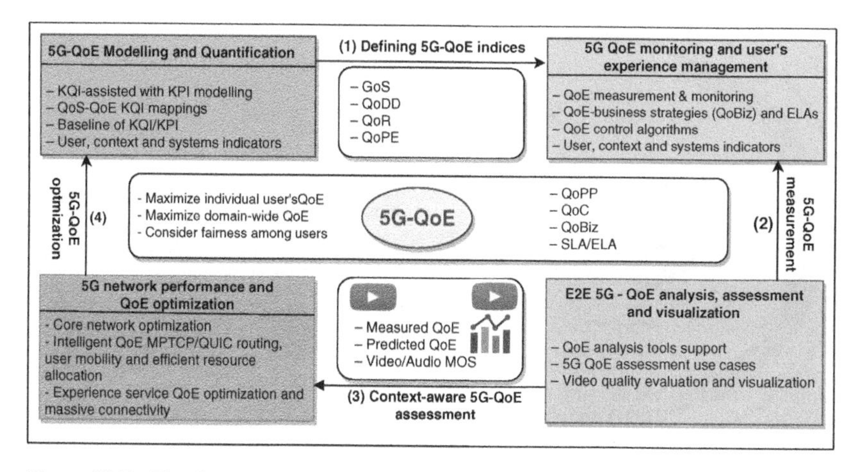

Figure 12.5 The QoE ecosystem on 5G and beyond networks.

elements of high-quality content using streaming codecs such as H.266/Versatile Video Coding (VVC), High-Efficiency Video Coding (HEVC/H.265), VP9, or AV1. The Service Provider (SP) ensures that the multimedia content is delivered to customers based on the Experience-Level Agreements (ELAs) [16] and Quality of Business (QoBiz) contracts [2]. Network providers have then to offer QoE-based connectivity services and reachability between network hosts. Finally, the customer should receive QoE-based services at anytime, anywhere to access services from SPs. The proposed 5G ecosystem for QoE provisioning consists of four parts namely the *Customer QoE Quantification, Customer Experience and Service Quality Management, an E2E QoE Analysis, Assessment, and Visualization,* and *5G Network Performances and Optimization.* It is centered on QoE as the driving performance metric of 5G ecosystem. We include the definition of QoE with key concepts for quality improvement and important QoE defining features as shown in Figure 12.6.

QoE encompasses the QoS, Grade of Services (GoS), Quality of Resilience (QoR), Quality of Design and Delivery (QoDD), Quality of Presentation and Perception (QoPP), QoBiz, and Quality of Conformance (QoC). The GoS is used to categorize 5G data services that have high levels of requirements as defined in the QoE offered through well-defined SLAs/ELAs. For each 5G network data service category, a class of service has to be defined and established for users. The QoPP is mainly related to the spatial content problems as perceived by the end users on the application level. The QoPP should be the main factor to be analyzed in the QoE models since it is used to provide feedback to users at the application layer. The QoDD is related to the network design below the application level with efficient capabilities to deliver data in time and therefore meeting the reliability,

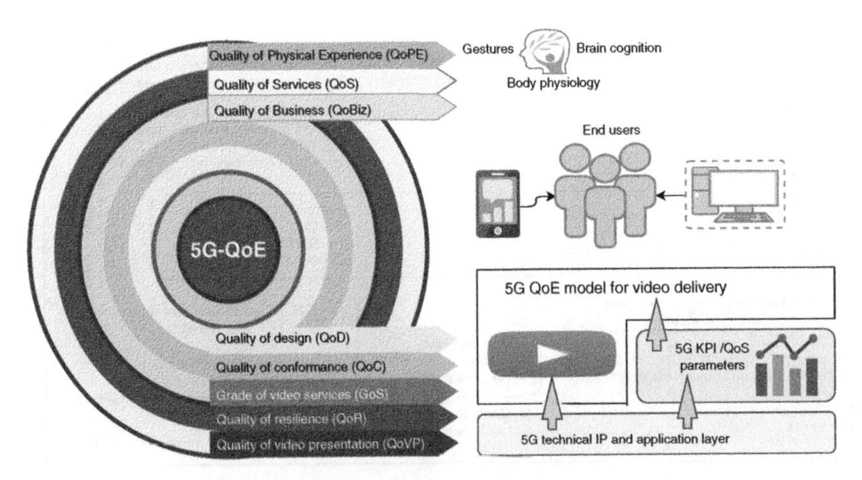

Figure 12.6 An expanded definition of QoE over 5G and beyond networks.

scalability, and flexibility which are among the strongest aspects of 5G softwarized and beyond networks.

During QoE cross-layer management design both QoPP and QoDD should be taken into account to provide an acceptable and the best possible user's QoE level. These two factors have to be considered because the video quality have to be created (video codecs), delivered/transmitted (at less packet loss, minimal delay, low latency, etc.), presented at the user's device (with minimum bitrate switching, low buffering events, no/less stalling events, etc.,) so that it can be perceived well by the end users. The proposed ecosystem also integrates the QoC (what the quality that a customer needs) to ensure that the delivered services through the QoE provisioning chain and all involved stakeholders meet the design specifications of 5G and beyond networks. We provide a description of components in the QoE provisioning ecosystem in 5G networks as follows:

12.4.1 Customer QoE Quantification

The customer QoE quantification defines the 5G service management guidelines and service quality design principles. It clarifies the customer perceptive and cognitive characteristics with respect to the service quality. It establishes a methodology for estimating QoE using quality-related information from the network, servers, and end-user's terminals. This step identifies and defines the KPI/KQI relationships as well as QoS to QoE mappings useful for establishing the key customer QoE requirements. The KPIs are quantifiable measurements that provide reflection of critical successful or unsuccessful factors of multimedia streaming services. Examples of KPIs are packet loss rate, delay, jitter, handover success rate, etc. The KQI defines the quality of a service perceived subjectively by the user. Examples of KQIs include service connect time, response time, availability, speech quality, etc. The KQI provides an indicator for a specific performance aspect of the product and draws their data from a number of sources including KPIs [17]. It is worth mentioning that the KPIs provide information related to the monitored resources while the KQIs are used for estimating the E2E QoE as perceived by the end users. The KPIs that are related to each of the KQIs are then defined and used to determine the service quality provided by network providers through the different measurement platforms and tools.

12.4.2 QoE Analysis, Assessment, Measurement, and Visualization

To help network operators and SPs to quickly solve the identified problems in 5G networks through monitoring, service developers, and experts conduct an E2E QoE assessment, service quality detection, and delimitation. At this point, network operators and SPs can perform network optimization tasks, 5G network performance measurements according to relevant service features from different network administrative domains. Based on recent advances in effective computing and sensing, Dudin et al. [18] propose an SDN/NFV-based resource allocation approach with automated QoE assessment in 5G/B5G wireless systems. Using a mobile Internet experience from end user perspective in Germany, Schwind et al. [19] propose a crowdsourced network measurements methodology that conducts more than 2.5 million throughput tests from Tutela Ltd, an independent crowdsourced data company with a global panel of over 300 million smartphone users. Recent work shows that crowdsourcing can support vendors, operators, and regulators to determine the end users in new 5G networks architecture that enable various new applications and network services [20].

12.4.3 5G Network Performance and QoE Optimization

From network operators' perspectives, the QoE-driven optimization involves network resource management mechanisms that maximize the end users' QoE. Such approaches are normally based on the quality-related information collected from the QoE monitoring and measurements process to provide quality assurance and service control in the network. Following the QoE quantification, monitoring, QoE analysis, assessment, and visualization, the next step is to perform QoE optimization (arrow 4 in Figure 12.5). QoE-enforcement and resource management can be achieved on the available information in the network (e.g. operator's QoE charging policy, service priority, available network resource, or user's profile and subscriber's data) by making optimal resource allocation decisions. Based on the specific QoE targets for a service as defined at the network-level QoE parameters, the QoE-driven management strategies can manage resources dynamically in the network.

12.4.4 Multistakeholders Collaboration for Video Content Delivery and Network Infrastructures

The value chain for video content distribution by involving multistakeholder collaboration is indicated in Figure 12.7 where the Content Owner sells its content to online Content Providers. ISPs' primary goal is to offer enough bandwidth to meet consumer demand for data. New bandwidth bottlenecks develop as demand rises, leading to additional performance degradation points. ISPs thus oversee a continual cycle of network expansion, modernization, and upkeep. ISP networks are under the greatest strain from video traffic, which is quickly exposing new bottlenecks and performance problems. ISPs have a strong incentive to make sure that their networks meet customer expectations as providers of competitive broadband services. ISPs must overcome a number of bandwidth-related obstacles while

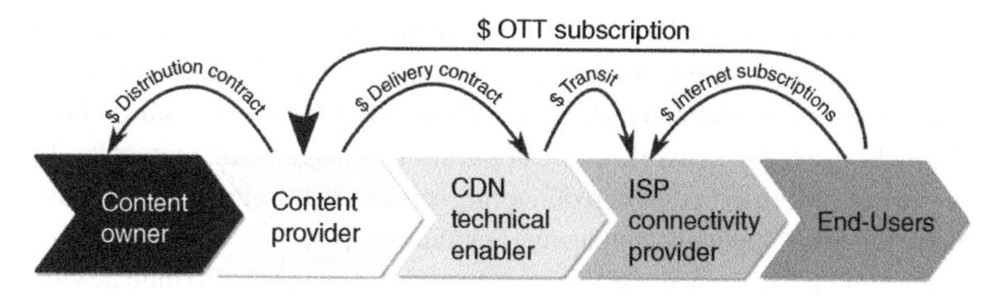

Figure 12.7 Multistakeholder value chain for video content distribution.

controlling network supply. The "last mile" services are purchased by many ISP networks from a national or local access network operator. Some ISPs have complete autonomy and create their own end-to-end networks. However, each ISP must deal with varied bottleneck dynamics depending on the level of investment made over time in various network components and the relative rate at which consumers adopt bandwidth-hungry applications.

ISPs observe how OTT operators are utilizing various video distribution models. Others engage in their own video edge delivery infrastructure while some invest in public CDN relationships. While OTT operators with densely populated (i.e. national) audiences and frequent streaming patterns are building private video network capacity by working directly with a relatively small number of ISPs, OTT operators with distributed (i.e. multinational) audiences or relatively infrequent streaming patterns continue to focus on their public CDN partnerships. The "edge" of the video delivery network is a hot topic in the distribution of video. This frequently refers to a CDN's edge, which subsequently communicates with the ISP network via peering or direct interconnection. ISPs then oversee a variety of broadly used network infrastructure to deliver the content to the user. The ISP's own data center infrastructure for data aggregation, the long-haul networks connecting all sites, their own CDN for on-net video distribution (if applicable), and the consumer premises equipment (CPE), including the home gateway, make up the infrastructure that receives the highest priority from ISPs.

In order to overcome QoE issues, OTT operators are concentrating on private CDN deployments within ISP networks. This plan aims to enhance the caliber of video delivery while safeguarding the experience of the final user. This investment can help the ISP ease the burden on peering points while also improving the OTT video viewing experience for their broadband subscribers. The end user's experience is crucial for both the ISP and the OTT operator as live video streaming expands and key events draw ever-larger OTT audiences.

Due to the increasing Internet video traffic, multistakeholders collaboration for video contents delivery and the overall QoE-based service management has been considered as a promising approach to overcome some of the peer-to-peer and CDN challenges [21]. Delivering multimedia services with high QoE to the end users is difficult when both ISPs and OTTs act independently. This is because QoE is a function of several parameters from ISPs (e.g. QoS) and OTTs (e.g. application parameters, context, and human factors). For monitoring the network status, OTT normally make use of the network-aware application management parameters. To manage the network resources when delivering services to the end users, ISPs use the application-aware network management strategies [22].The collaboration for multimedia QoE-aware service management between ISPs and OTTs could be vital in increasing their profits while providing multimedia services with better quality to their end users.

From the perspective of multistakeholders collaboration, the proposed QoE provisioning ecosystem shown in Figure 12.5 can help content delivery systems to improve the efficiency of the video contents distribution and optimize the overall performance of 5G networks by using information provided by the network operators and the network characteristics [22]. The collaboration between ISP and CDN can provide them with triple-win benefits. For example, ISP can gain better QoE-traffic management and efficient network resources utilization leading to a reduced cost of operation and investment. The CDN can acquire the network information and use them to improve the performance of link load and user's experience based on location [23]. An SDN-based CDN–ISP collaboration solution can greatly improve the efficiency of content delivery and the end users' QoE. The CDN–ISP collaboration architectures in this direction are proposed in [24–26]. Akyildiz et al. [24] propose SoftAir, a software-defined architecture for 5G wireless networks that improves the efficiency of content delivery. The architecture adopts different approaches such as distributed and collaborative traffic classifiers and mobility-aware control traffic balancing to optimize the control of traffic control and network monitoring. Wichtlhuber et al. [26] propose an SDN-based architecture that provides a fine grained, integrated traffic engineering for CDN traffic in the ISPs network. The CDN provider is given an ability to decide on the selection of surrogate servers. Wang et al. [25] propose NetSoft, a software-defined decentralized mobile network architecture for 5G to improve the efficiency of content delivery in 5G systems.

The QoE-aware OTT–ISP collaboration in service management has been investigated in recent years for improving QoE of users. Floris et al. [27] propose a QoE-aware service management reference architecture for a possible collaboration and information exchange among OTT and ISP in terms of technical and economic aspects. A joint venture, customer lifetime value, and QoE-fairness mechanisms are proposed to maximize the revenue by providing better QoE to customers paying more. Authors in [28] propose a PPNet, a framework that enables isolation between the SP's web and video provider's interfaces. PPNet also allows CDN–ISP collaboration while preventing the ISP's access to the video request and availability

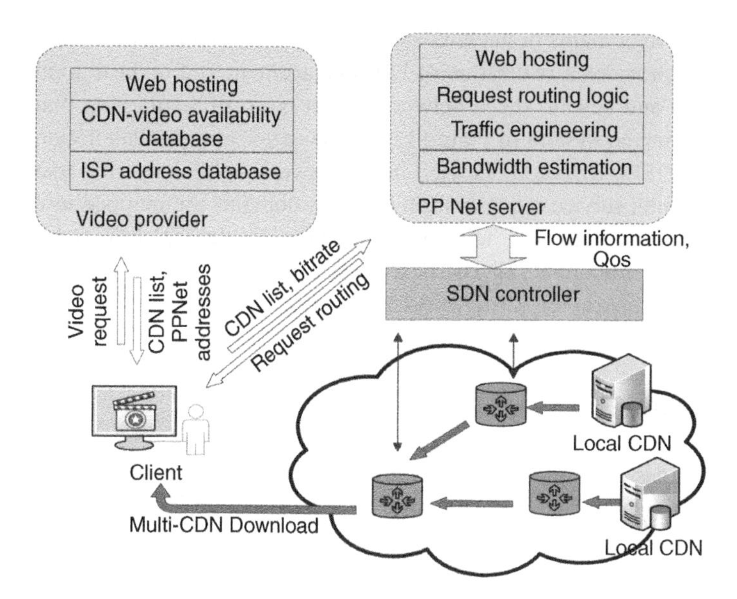

Figure 12.8 ISP–CDN collaboration and the elements of video delivery for 2030 and beyond networks.

information. An extended framework of PPNet in the context of 2030 and beyond networks is shown in Figure 12.8.

The goal of this architecture is to keep the ISP servers that provide network availability and the CDN servers that provide information on video availability apart. The CDN interface presents a server listing for each segment and quality level for the requested video, and the client establishes simultaneous connections to both interfaces. Without providing details about the movie itself, this information is sent to the ISP interface. The PPNet architecture lists the following as the active video delivery components: (i) SDN infrastructure with controller and switches placed in the ISP's network, (ii) local CDN nodes situated in the ISP's network, (iii) PPNet server, and (iv) the server of the video provider [26]. To answer to customer queries, the video supplier operates a web server. A mapping from potential client IP address ranges to ISP providers is stored on the server. It is possible to dynamically receive and update this list from ISPs who have a contract with the provider. The database of videos and quality settings accessible in each CDN node is likewise stored on the server. A mapping for the ISP company that hosts the CDN is present for each CDN node in the database. The cache replacement techniques that each CDN uses control the availability of videos within that CDN. While the ISP oversees their network, video providers handle the CDN nodes.

Ahmad et al. [29] propose a novel service delivery approach that is purely driven by the end user's QoE while considering the collaboration between OTT and ISP. Recent developments of network slicing for multidomain orchestration and management provides a realization of E2E management and orchestration of resources in 5G and beyond sliced networks [2]. It is worth mentioning that the QoE provisioning ecosystem in Figure 12.5 emphasizes on the collaboration between verticals (ISPs, CDNs, OTTs, etc.) in 5G networks while ensuring that multimedia streaming service requests from different domains are mapped into multioperator and multiprovider technology domains while matching each service ELA/SLA requirements [2]. The next subsection provides a multimedia streaming use-case scenario for 5G networks and beyond networks.

12.5 Conclusion

This chapter provides the multimedia streaming services delivery aspects in the 2030 and beyond networks. The chapter focuses on new capabilities, features of the communications networks and services required from 2030 and beyond networks. This chapter presents an OTT/CDN-ISP collaboration for QoS-aware multi-CDN adaptive video streaming in future 2030 and beyond networks. It also provides the blockchain-enabled multimedia streaming resource management framework in 2030 networks along with the general QoE provisioning ecosystem for 2030 and beyond networks. The chapter also presents the new paradigms of internetworking for 2030 and beyond networks including the next-generation human–machine interaction network which is a human-centric hyper-real experience. It is important to note that a new core element called QML will be created by combining QC and machine learning. This calls for the use of QML to accelerate the learning of all transactions occurring at the 6G Network's edge in order to automatically change the network's resources. The processing power available in QML will easily cope with the future increase of video traffic and heterogeneous data. QML's role in the 2030 communication network will be to optimize the video streaming traffic for the 2030 applications and services.

Bibliography

1 Giordani, M., Polese, M., Mezzavilla, M. et al. (2020). Toward 6G networks: use cases and technologies. *IEEE Communications Magazine* 58 (3): 55–61.

2 Barakabitze, A.A., Ahmad, A., Mijumbi, R., and Hines, A. (2020). 5G network slicing using SDN and NFV: a survey of taxonomy, architectures and future challenges. *Computer Networks* 167: 1–40.

3 Akyildiz, I.F., Kak, A., and Nie, S. (2020). 6G and beyond: the future of wireless communications systems. *IEEE Access* 8: 13399–134030.

4 Barakabitze, A.A., Barman, N., Ahmad, A. et al. (2019). QoE management of multimedia services in future networks: a tutorial and survey. *IEEE Communication Surveys and Tutorials.* 22 (1): 526–565.

5 Raake, A., Borer S., Satti S. M., et al. (2020). Multi-model standard for bitstream-, pixel-based and hybrid video quality assessment of UHD/4K: ITU-T P.1204. *Human-Centric Computing and Information Sciences* 10 (53): 1–27.

6 Prasad, R. (2016). Knowledge home. *International Conference on Advanced Computer Science and Information Systems (ICACSIS)*, pp. 33–38, June 2016.

7 Cianca, E., Sanctis, M.D., Mihovska, A., and Prasad, R. (2018). Communications, navigation, sensing and services (CONASENSE). *Journal of Communication, Navigation, Sensing and Services Journal Merged with Journal of Mobile Multimedia* March 2018.

8 Rufino Henrique, P.S. and Prasad, R. (2021). 6G networks for next generation of digital TV beyond 2030. *Wireless Personal Communications* 121: 1363–1378.

9 Velan, P., Cermăk, M., Celeda, P., and Drašar, M. (2015). A survey of methods for encrypted traffic classification and analysis. *International Journal of Network Management* 25: 355–374.

10 Clemm, A., Vega, M.T., Ravuri, H.K. et al. (2020). Modeling the time-varying subjective quality of HTTP video streams with rate adaptations. *IEEE Communications Magazine* 58 (1): 93–99.

11 Huang, T., Yang, W., Wu, J. et al. (2019). A survey on green 6G network: architecture and technologies. *IEEE Access* 121: 175758–175768.

12 Communication Networks 2030 (2021) Building a Fully Connected, Intelligent World. https://www-file.huawei.com/-/media/corp2020/pdf/giv/industry-reports/communications.network.2030.en.pdf (accessed 13 August 2022).

13 European Vision for the 6G Network Ecosystem. https://5g-ppp.eu/wp-content/uploads/2021/06/WhitePaper-6G-Europe.pdf (accessed 13 August 2022).

14 Xu, H., Klaine, P.V., Onireti, O. et al. (2020). Blockchain-enabled resource management and sharing for 6G communications. *Digital Communications and Networks* 6 (3): 261–269.

15 Herbaut, N. and Negru, N. (2017). A model for collaborative blockchain-based video delivery relying on advanced network services chains. *IEEE Communications Magazine* 55 (9): 70–76.

16 Varela, M., Zwickl, P., Reichl, P. et al. (2015). From service level agreements (SLA) to experience level agreements (ELA): the challenges of selling QoE to the user. *In proceedings of IEEE ICC QoE-FI*, June 2015.

17 Vaser, M. and Forconi, S. (2016). QoS KPI and QoE KQI relationship for LTE video streaming and VoLTE services. *9th International Conference on Next Generation Mobile Applications, Services and Technologies*, Volume 7, pp. 318–323, January 2016.

18 Dudin, B., Ali, N.A., Radwan, A., and Taha, A.-E.M. (2019). Resource allocation with automated QoE assessment in 5G/B5G wireless systems. *IEEE Network* 33 (4). 76–81

19 Schwind, A., Wamser, F., Hoßfeld, T. et al. (2020). Crowdsourced Network Measurements in Germany: Mobile Internet Experience from End User Perspective. *https://arxiv.org/abs/2003.11903*.

20 Tobias, H. et al. (2020). White paper on crowdsourced network and QoE measurements - definitions, use cases and challenges. *Seminar on Crowdsourced Network and QoE Measurements*, pp. 1–24, March 2020. https://doi.org/10.25972/OPUS-20232.

21 Akpinar, K. and Hua, K.A. (2020). PPNet: Privacy protected CDN-ISP collaboration for QoS-aware multi-CDN adaptive video streaming. *ACM Transactions on Multimedia Computing, Communications, and Applications* 16 (49): 1–23.

22 Ahmad, A. and Atzori, L. (2020). MNO-OTT collaborative video streaming in 5G: the zero-rated QoE approach for quality and resource management. *IEEE Transactions on Network and Service Management* 17 (1): 361–374.

23 Jia, Q. et al. (2017). The collaboration for content delivery and network infrastructures: a survey. *IEEE ACCESS* 5: 18088–18106.

24 Akyildiz, I.F. et al. (2015). SoftAir: A software defined networking architecture for 5G wireless systems. *IEEE Network* 85: 1–18.

25 Wang, H. et al. (2015). SoftNet: A software defined decentralized mobile network architecture toward 5G. *IEEE Network* 29 (2): 16–22.

26 Wichtlhuber, M. et al. (2015). An SDN-based CDN/ISP collaboration architecture for managing high-volume flows. *IEEE Transactions on Network and Service Management* 12 (1): 48–60.

27 Floris, A. et al. (2018). QoE-aware OTT-ISP collaboration in service management: architecture and approaches. *ACM Transactions on Multimedia Computing, Communications, and Applications* 14 (2): 1–24.

28 Akpina, K. and Hua, K.A. (2020). PPNet: Privacy protected CDN-ISP collaboration for QoS-aware multi-CDN adaptive video streaming. *ACM Transactions on Multimedia Computing, Communications, and Applications* 16 (2): 1–23.

29 Ahmad, A. et al. (2018). QoE-centric service delivery: a collaborative approach among OTTs and ISPs. *Computer Networks* 110: 168–179.

Index

Note: Page numbers in *italics* denote figures and page numbers in **bold** denote tables.

Multimedia Streaming in SDN/NFV and 5G Networks: Machine Learning for Managing Big Data Streaming, First Edition. Alcardo Barakabitze and Andrew Hines.

i

o

p

q

Printed and bound by CPI Group (UK) Ltd, Croydon, CR0 4YY

16/04/2025

14658579-0002